OUT OF PRINT

Introduction to Multivariate Analysis

Introduction to Multivariate Analysis

CHRISTOPHER CHATFIELD
ALEXANDER J. COLLINS
School of Mathematics
Bath University

LONDON NEW YORK
CHAPMAN AND HALL

First published 1980 by
Chapman and Hall Ltd
11 New Fetter Lane, London EC4P 4EE
Reprinted 1983, 1986

Published in the USA by
Chapman and Hall
29 West 35th Street, New York NY 10001

Printed in Great Britain at the
University Press, Cambridge

ISBN 0 412 16030 7 (cased)
ISBN 0 412 16040 4 (paperback)

British Library Cataloguing in Publication Data

Chatfield, Christopher
Introduction to multivariate analysis
1. Multivariate analysis
I. Title II. Collins, Alexander J
519.5′3 QA278 80–40603

ISBN 0–412–16030–7
ISBN 0–412–16040–4 Pbk

Contents

PART TWO: FINDING NEW UNDERLYING VARIABLES

PART THREE: PROCEDURES BASED ON THE MULTIVARIATE NORMAL DISTRIBUTION

PART FOUR: MULTIDIMENSIONAL SCALING AND CLUSTER ANALYSIS

Preface

This book provides an introduction to the analysis of multivariate data. It should be suitable for statisticians and other research workers who are familiar with basic probability theory and elementary inference, and also have a basic grounding in matrix algebra. The book should also be suitable as a text for undergraduate and postgraduate statistics courses on multivariate analysis.

The book covers a wider range of topics than some other books in this area. It deals with preliminary data analysis, principal component and factor analysis and traditional normal-theory material. It also covers cluster analysis and scaling techniques.

In writing the book, we have tried to provide a reasonable blend of theory and practice, in contrast to much of the existing literature which concentrates on one or other aspect. Enough theory is given to introduce the concepts and to make the topics mathematically interesting. But we also discuss the use (and misuse) of the techniques in practice and present appropriate real-life examples.

Although the book is primarily an introductory text, we have nevertheless added appropriate references for further reading, so that the reader may extend his studies if he wishes.

The book is divided into four parts with self-explanatory titles. One of us (C.C.) was primarily responsible for Parts I, II and IV, while the other (A.J.C.) was primarily responsible for Part III. We recommend that all readers start with Part I, and Chapter 4 of Part II, but the reader may then go on directly to Parts III or IV or Chapter 5 of Part II. The book contains more material than can be covered in a typical introductory lecture course and we suggest that a selection of topics be made from Chapters 5–11, depending on the aims of the course and the interests of the students.

The advent of high-speed computers has revolutionized the subject in recent years and it is now possible to carry out complicated analyses on large-scale data sets very easily. This brings dangers as well as benefits. The reader should beware of using techniques on inappropriate data, and of using 'poor' computer package programs. However, we hesitate to make recommendations on computer packages as the situation changes so rapidly.

We would like to thank A. Bowyer, W. Lotwick, C. Osmond, and R. Sibson for computational help in producing the examples of Chapters 10 and 11. We are also indebted to many people for helpful comments on earlier drafts of this book, including M. A. Collins, D. R. Cox, G. J. Goodhardt, S. Langron, W. Lotwick, P. K. Nandi, C. Osmond and R. Sibson.

Of course any errors, omissions or obscurities which remain are entirely our responsibility. The authors will be glad to hear from any reader who wishes to make constructive comments.

School of Mathematics
Bath University
Bath, Avon
England
January 1980

CHRIS CHATFIELD
ALEC COLLINS

1986 REPRINT

Some references have been updated in this reprint. In addition a few mistakes have been corrected, notably one arithmetical error in Table 8.7 which also affected later results.

We thank readers who have made constructive comments, especially G. M. Arnold.

June 1986

PART ONE
Introduction

The first three chapters contain a variety of important introductory material.

Chapter 1 begins with some typical multivariate problems, and discusses ways of tackling them. The chapter also includes a section revising the more important results in matrix algebra.

Chapter 2 introduces the reader to multivariate probability distributions.

Chapter 3 shows how to carry out a preliminary analysis of a set of data. In particular, the chapter discusses how to feed the data into a computer without making too many errors, and also discusses the calculation of summary statistics such as means, standard deviations and correlations.

CHAPTER ONE
Introduction

Multivariate data consist of observations on several different variables for a number of individuals or objects. Data of this type arise in all branches of science, ranging from psychology to biology, and methods of analysing multivariate data constitute an increasingly important area of statistics. Indeed, the vast majority of data is multivariate, although introductory statistics courses naturally concentrate on the simpler problems raised by observations on a single variable.

1.1 Examples

We begin with some examples of multivariate data.

(*a*) *Exam results.* When students take several exams, we obtain for each student a set of exam marks as illustrated in Table 1.1. Here the 'variables' are the different subjects and the 'individuals' are the students.

The analysis of data of this type is usually fairly simple. Averages are calculated for each variable and for each individual. The examiners look at the column averages in order to see if the results in different subjects are roughly comparable, and then examine the row averages in order to rank the individuals in order of merit. This ranking is usually achieved by simply ordering the average marks of the students, though some more complicated averaging procedure is sometimes used. If the results for one exam appear to be out of line with the remainder, then these marks may be adjusted by the examiners. For example, in Table 1.1 the mathematics average is somewhat

Table 1.1 Some typical exam results

Name	History	Mathematics	Physics	Geography	Average
Adams, E.	58	43	45	61	51.7
Bloggs, J.	43	58	59	48	52.0
Fisher, R.	37	35	43	52	41.7
etc.					
Average	48.2	43.7	50.2	47.2	47.3

low and it might be deemed fair to scale the mathematics marks up in some way.

Although the above analysis is fairly trivial, it does illustrate the general point that multivariate analyses are often concerned with finding relationships, not only between variables, but also between individuals.

A more sophisticated analysis might try to establish how particular students did particularly badly or particularly well, and also see if results in different subjects are correlated. For example, do students who perform above average in science subjects tend to perform below average in arts subjects (i.e., is there negative correlation)?

(*b*) *A nutritional study.* A survey was recently carried out on children in Nepal to examine the use of three measured variables in assessing nutritional status. The variables were height, weight and middle-upper-arm circumference (MUAC for short). The data also recorded the sex of each child (coded as 1 for males and 2 for females), the age of each child, and the social caste of each child (coded from 1 to 6). A very small portion of the data is shown in Table 1.2.

Note that some of the variables are continuous (e.g., height) while others are discrete (e.g., caste).

The analysis of the results, which will not be described here, was concerned with several aspects of nutrition. First, what are the relationships, if any, between the variables? Second, what differences, if any, are there between the castes? Third, how can one assess acute undernutrition?

Table 1.2 Part of the Nepal nutritional data

Code number of child	Sex	Caste code	Age (months)	Height (cm)	Weight (kg)	MUAC (cm)
1	1	1	52	92	12.1	15.0
2	1	1	14	72	8.7	14.0
3	1	1	52	83	13.5	18.0
4	1	5	27	77	9.7	15.0
5	2	4	32	89	11.6	14.5
etc.						

Table 1.3 Part of the spina bifida data

Code number of town	Number of births	Spina bifida deaths	Total deaths	SEI	Elements (parts per million)			
					Ca	Mg	Na	Cu
1	3191	4	64	123	97.0	5.2	16.2	0.35
2	5292	19	98	106	16.2	1.8	5.8	0.05
3	5898	13	116	91	12.4	1.8	5.2	0.03
4	5005	3	83	95	118.0	9.8	44.6	0.48
etc.								

(*c*) *Medical data.* Some doctors suspect that the number of spina bifida cases in babies is related to social conditions and to the composition of tap water. Various towns in England and Wales were selected in a recent study, and many tests were made on samples of water to assess the amount of trace elements present. An index of socio-economic status (the SEI) was also found for each town. The number of spina bifida deaths in one year was recorded along with the total number of births and the total number of deaths before age 1 from all causes. A small portion of the data is shown in Table 1.3.

Many questions were asked in the analysis of these data. First, we notice that there are large variations in the amount of trace elements in tap water, and also that there appears to be substantial variation in the number of spina bifida deaths. Can we explain the variation in spina bifida deaths (and in total deaths) in terms of some function of the trace elements and the SEI?

Here we note that the variables are not of the same type. In the jargon of multiple regression, we say that 'spina bifida deaths' and 'total deaths' are *dependent* variables, while SEI and the proportions of trace elements are *explanatory* variables, in that they are used to 'explain' the variation in the dependent variables. Note that explanatory variables are sometimes called *predictor*, *controlled* or *independent* variables. The latter term is particularly inappropriate for economic data where the explanatory variables are often highly correlated.

(*d*) *Other examples.* Many other examples could be given. For example, archaeologists have studied various burial sites and noted whether certain features, called artifacts, are present (e.g., whether or not pottery containers have handles). From this they hope to date the burial sites and in particular to arrange them in temporal sequence. Another classic study in biometrics/archaeology was concerned with the skulls found in an ancient burial ground. Given measurements on the shape of the skulls, the main problem was to see if all the skulls belonged to members of the same race, or if they came from two different races as might happen if the deaths resulted from a battle.

Various other examples will be introduced later in the book. These include a study of the variation in certain types of criminal offences (see Sections 3.2 and 4.5).

1.2 Notation

We denote the number of variables by p, and the number of individuals or objects by n. Thus we have a total of $(n \times p)$ measurements. In future we will usually refer to 'individuals', rather than 'individuals or objects', in order to save space.

Let

$$x_{rj} = r\text{th observation on the } j\text{th variable}$$
$$(r = 1, \ldots, n; j = 1, \ldots, p)$$

The matrix whose element in the rth row and jth column is x_{rj}, will be called the *data matrix* and will be denoted by X. Thus

$$X = \begin{bmatrix} x_{11} & x_{12} & \dots & x_{1p} \\ x_{21} & & \dots & x_{2p} \\ \vdots & & & \\ x_{n1} & & \dots & x_{np} \end{bmatrix}$$

Note that some authors (e.g. Anderson, 1958; Kendall, 1975) reverse the order of the subscripts in the above definition of x_{rj}, though our definition is becoming increasingly popular (e.g., Morrison, 1976; Maxwell, 1977). But the reader obviously needs to check which definition is being used whenever he refers to another book or journal.

There is a practical reason why the data matrix should have n rows (and p columns) rather than n columns (and p rows). The value of n is usually much bigger than the value of p, and the obvious way to record the data is by means of a row-vector for each individual.

Indeed, the data matrix can be seen as n row vectors, which we denote by $\mathbf{x}_1^T$ to $\mathbf{x}_n^T$, or as p column vectors, which we denote by $\mathbf{y}_1$ to $\mathbf{y}_p$. Thus

$$X = \begin{bmatrix} \mathbf{x}_1^T \\ \mathbf{x}_2^T \\ \vdots \\ \mathbf{x}_n^T \end{bmatrix} = [\mathbf{y}_1, \mathbf{y}_2, \dots, \mathbf{y}_p]$$

where $\mathbf{x}_i^T$ denotes the *transpose* of $\mathbf{x}_i$ (see Section 1.6). Note that vectors are printed in bold type, but we depart from established convention by printing matrices in ordinary type. This not only reduces typesetting costs, but also allows the useful distinction between vectors and matrices in many equations.

The choice of r and j as the two subscripts in x_{rj} is deliberately made so as to distinguish clearly between the subscripts for the variable and for the individual. If the observation is denoted by x_{ij}, as it is in some books, it is not clear which subscript refers to the variable and which to the individual. More generally, we use $i, j, \dots$ as subscripts for variables, and $r, s, t, \dots$ as subscripts for individuals. Thus when comparing two variables, we will use i and j as subscripts. For example, in Chapter 2 the correlation coefficient for variables i and j is defined, and is denoted by ρ_{ij}. Alternatively, when comparing two individuals, we typically use r and s as subscripts. For example, in Chapter 10 we denote the 'distance' between individuals r and s by d_{rs}.

1.3 Review of objectives and different approaches

A wide variety of multivariate techniques are available. The choice of the most appropriate method depends on the type of data, the type of problem, and the sort of objectives which are envisaged for the analysis. The underlying theme of much multivariate analysis is *simplification*. In other words, we want to summarize a large body of data by means of relatively few parameters. We also note that many multivariate techniques are *exploratory* in that they seek to generate hypotheses rather than test them.

One fundamental distinction between the many techniques is that some analyses are primarily concerned with relationships between *variables*, while others are primarily concerned with relationships between *individuals*. Techniques of the former type are called *variable-directed*, while the latter are called *individual-directed*.

In the previous section we pointed out that the data matrix can be regarded as n row rectors or as p column vectors. The row vectors are points in p-dimensional space, while the column vectors are points in n-dimensional space. When comparing variables, we compare the column vectors. In Chapter 3 we will see that one way of comparing variables is to calculate the correlation coefficient. This is essentially the dot product of the two column vectors after standardizing them so that the observations in each column have zero mean and unit variance. In fact it can be shown that the correlation coefficient is the cosine of the angle between the two standardized n-dimensional column vectors. For example, if the two variables are 'close together' in n-dimensional space, then the angle between the two vectors will be 'small' and its cosine will be close to $+1$, indicating 'high' positive correlation.

In order to compare two individuals, we compare the corresponding row vectors in p-dimensional space. Various ways of assessing the 'distance' between two individuals are discussed in Chapter 10.

The distinction between an analysis in p-space and in n-space has already been made in Example (*a*) of Section 1.1. In assessing exam results, we not only want to compare the results of different students, but also want to compare the standard of different exams.

Analyses based on the $(p \times p)$ matrix of covariances or correlations between variables are sometimes called R-techniques by non-statisticians, while analyses based on the $(n \times n)$ matrix of 'distances' between individuals are sometimes called Q-techniques. We shall not use this terminology.

We should add that sometimes the distinction between individual-directed and variable-directed techniques is not clear-cut. For example, it is sometimes useful to look at the variables and create one or more new variables by an appropriate transformation in order to compare individuals more easily.

Analysis of interdependence. Suppose we have a set of variables which arise 'on an equal footing', as for example when we measure different dimensions of different members of a particular species. If our primary interest is in the variables, we are interested in analysing the interdependence of the variables. At one extreme, the variables may be completely independent, while at the other extreme, there may be collinearity, with one variable being a linear function of the other variables.

With just two variables, we simply look at the correlation between them.

With more than two variables, a technique called *principal component analysis* may be appropriate (see Chapter 4). This technique aims to transform the observed variables to a new set of variables which are uncorrelated and arranged in decreasing order of importance. The principal aim is to *reduce the dimensionality* of the problem and to find new variables which make the data easier to understand.

Factor analysis (see Chapter 5) has somewhat similar aims but assumes a proper statistical model which takes explicit account of error. This method is controversial in many of its applications.

One may also ask if the variables fall into groups so that there is high correlation between variables in the same group, but low correlation between groups. This question is sometimes answered using factor analysis, but it is often easier to answer by just looking at the correlation matrix.

Analysis of dependence. In Example (*c*) of Section 1.1, the variables 'do not arise on an equal footing'. The use of this phrase is not meant to imply that some variables are more important than others (though they may be). Rather it implies that there are dependent variables and explanatory variables.

In multiple regression we try to explain the variation in one dependent variable by means of the variation in several explanatory variables. In multivariate regression (see Chapter 9) we have more than one dependent variable. The study of functional and structural relationships is also relevant.

Classification. An individual-directed problem is to ask if the individuals form groups or clusters, or if they are more or less randomly scattered over the domain of variation. Techniques for forming groups of individuals are usually called *cluster analysis* (see Chapter 11).

A different type of classification problem arises in *discriminant analysis.* Suppose we have a random sample of individuals from population A and another sample from population B (i.e., we have two known 'clusters'). How do we set up a rule to allocate new individuals, whose origin is unknown, to the correct population with the greatest probability of being correct? The allocation rule which is found (see Chater 7) usually depends on forming a new variable which is a linear function of the observed variables. One obvious example is in medicine, where a number of symptoms (presence or absence) are recorded for patients who are suspected of having a certain disease. Later tests (perhaps after death!) indicate for certain whether they have the disease or not. From these results we can work out which symptoms are most discriminatory in making a diagnosis for a new patient.

Other approaches. Many other types of problem may arise. For example, another individual-directed objective is *ordering* (rather than clustering), where one wants to rank the 'individuals' in some sort of order. Another class of methods is *scaling* (see Chapter 10), which is an individual-directed technique that is appropriate when the data are in the form (or can readily be put in the form) of similarities or dissimilarities between individuals. The main objective of scaling is to produce a 'map' or configuration of the individuals in a small number of dimensions so as to compare the individuals more easily. In *classical scaling* (sometimes called *principal co-ordinates analysis*), the aim is an algebraic reconstruction of the positions of the individuals assuming that the distances are approximately Euclidean. An alternative scaling technique, often called *non-metric multidimensional scaling*, has a similar aim but only uses the rank order of the distances, and so we prefer to use the description *ordinal scaling*.

1.4 Some general comments

(*a*) Some of the techniques we shall meet are generalizations of familiar methods such as correlation and regression. Others are completely new (e.g., principal component analysis).

(*b*) As in the univariate case, our data may be a sample from an infinite population, from a finite population, or they may form a complete census. In the latter case there are no sampling or inference problems. For example, if we have data on certain features of all British parliamentary constituencies, we would simply try to summarize the data.

(*c*) The computer has revolutionized the subject and led to a large expansion in the number of analyses being undertaken. But computer package programs are not all they might be, and it is very easy for them to be used on inappropriate data or in an inappropriate situation.

(*d*) Getting the data on to the computer is not easy and errors are likely to occur (see Section 3.1). This is particularly true in the social sciences, where data are inclined to be 'messy'.

(*e*) Plotting the data is not easy when they are more than two-dimensional (see Section 3.3).

(*f*) Some multivariate procedures, such as cluster analysis and ordinal scaling, are essentially distribution-free methods (or non-parametric methods). But there are difficulties in finding multivariate analogues of some univariate non-parametric procedures based on order statistics. Whereas it is easy to order (or rank) univariate observations, there is no natural basis for ordering multivariate data. Barnett (1976) discusses this problem and describes some restricted forms of ordering.

(*g*) Many multivariate methods are based on the assumption that the data have an approximate multivariate normal distribution. The main reason for this seems to be that the multivariate normal distribution is much easier to handle mathematically than other multivariate distributions, and because distribution-free methods are not always available in the multivariate case (see (*f*) above). Some data can be transformed closer to normality, but it is still common to find the normal assumption being made for data which are nowhere near normal. Fortunately, the methods based on the normal assumption are reasonably 'robust' to departures from normality.

(*h*) It is useful to distinguish between the different types of variables which arise in different situations. Sometimes the variables are spatial co-ordinates in two or three dimensions. These are what one might call genuine multidimensional variables. The choice of origin and the orientation of the axes is clearly arbitrary.

A second type of situation is where the variables are all of the same type. For example, they might all be dimensions of an animal, or all percentages of different elements in a compound, or all scores on a 5-point rating scale. Such variables are directly comparable, and often have similar variances.

A more complicated situation arises when the variables are of different types, as in Examples (*b*) and (*c*) of Section 1.1. Some variables may be discrete and some continuous, while some may have much larger variances than others. Then the choice of metric may be crucial. One way of proceeding is to standardize the values so that each variable has zero mean and unit variance. Thus one works with correlations rather than covariances. It is rather disturbing that many multivariate techniques give quite different answers if one works with the raw data rather than with the standardized values (e.g., see Section 4.4).

The treatment of continuous variables is often different from that of discrete variables, particularly when the latter are *binary* (i.e., only two possible values). The treatment of *qualitative* data is discussed briefly in Sections 3.1 and 3.4

1.5 Review of books on multivariate analysis

This section gives a brief review of some other books on multivariate analysis. These books will complement our own book by providing an alternative, and in some cases more extended, view of multivariate topics.

The books by Morrison (1976), Mardia *et al.* (1979) and Kendall (1975) are perhaps the most suitable additional choices for the beginner, though Morrison does not discuss scaling or cluster analysis, and Kendall's notation is not always easy to understand. The classic text by Anderson (1958) is based entirely on the multivariate normal distribution and is suitable for the more mathematical reader. Marriott (1974) provides a brief overview of the area with little detail. Seal (1964) is aimed mainly at non-specialists and covers the general linear model, principal component analysis, canonical analysis and factor analysis in a fairly readable way. The book by Gnanadesikan (1977) contains a lot of useful ideas, including plotting techniques, for reducing dimensionality and clustering. Cooley and Lohnes (1971) is written for 'data-analysts'. Everitt (1978) concentrates on graphical techniques.

Many books have been written for specialists in other areas, such as psychologists and biologists. Blackith and Reyment (1971) is aimed primarily at biologists and contains much useful discussion of multivariate techniques but virtually no mathematics. Maxwell (1977) is aimed at behavioural scientists but appears rather concise for beginners.

Many more books have been written on one particular aspect of multivariate analysis, including Everitt (1974) on cluster analysis, Lawley and Maxwell (1971) on factor analysis and Everitt (1984) on latent variable models.

1.6 Some matrix algebra revision

Throughout this book, we assume that the reader is familiar with standard operations on vectors and matrices. For reference purposes, this section briefly revises some of the definitions and results which are particularly useful in multivariate analysis. Further details are given by Morrison (1976, Chapter 2).

A *matrix* is a rectangular array of elements. If it has m rows and n columns, it is said to be of order $(m \times n)$. The element in the ith row and jth column of a matrix A will be denoted by a_{ij}. When multiplying two matrices together, it is often useful to write their order underneath to check that the numbers of rows and columns agree. Thus if B is of order $(n \times p)$, then

$$\underset{(m \times n)}{A} \times \underset{(n \times p)}{B} = \underset{(m \times p)}{C}$$

Remember that multiplication is associative, so that $A(BC) = (AB)C$; but *not* commutative, so that in general $AB \neq BA$.

The *transpose* of a matrix A is obtained by interchanging rows and columns and will be written A^{T}. A useful result is that

$$(AB)^{\mathrm{T}} = B^{\mathrm{T}} A^{\mathrm{T}} \tag{1.1}$$

A square matrix A is said to be *symmetric* if $A = A^{\mathrm{T}}$, so that $a_{ij} = a_{ji}$ for every i and j. Many symmetric matrices arise in multivariate analysis.

The *trace* of a square matrix A of order $(p \times p)$ is the sum of the diagonal terms, namely $\sum_{i=1}^{p} a_{ii}$. It can be shown that

$$\operatorname{trace}(AB) = \operatorname{trace}(BA) \tag{1.2}$$

provided (AB) is square, though A and B do not have to be square.

A square matrix is said to be *diagonal* if all its off-diagonal elements are zero. The *identity* matrix, denoted by I, is a diagonal matrix whose diagonal elements are all unity. If I has q rows (and hence q columns), it is said to be of order q and is often denoted by I_q.

The *null* matrix consists entirely of zeroes, and is denoted by 0 whatever its order.

The *determinant*, $|A|$, of a square matrix A of order $(p \times p)$ is a scalar, such that

$$|A| = a_{11}A_{11} + a_{12}A_{12} + \ldots + a_{1p}A_{1p}$$

where $A_{ij} = (-1)^{i+j} \times$ determinant of the matrix without the ith row and the jth column.

The determinant is unchanged if equal multiples of the elements of one row (or column) are added to the corresponding elements of any other row (or column). The determinant changes sign if any two rows (or any two columns) are interchanged. A matrix is said to be *non-singular* provided that $|A| \neq 0$. It can be shown that if A and B are square matrices of the same order, then

$$|AB| = |A|\,|B| \tag{1.3}$$

It can also be shown that if a square matrix A is written as

$$A = \begin{bmatrix} A_{11} & A_{12} \\ A_{21} & A_{22} \end{bmatrix}$$

where A_{11} is square and non-singular, then

$$|A| = |A_{11}| \times |A_{22} - A_{21}A_{11}^{-1}A_{12}| \tag{1.4}$$

In Equation (1.4), A_{11}^{-1} denotes the *inverse* of matrix A_{11}. The inverse of a square matrix A is a unique matrix A^{-1} such that

$$AA^{-1} = A^{-1}A = I \tag{1.5}$$

A necessary and sufficient condition for A^{-1} to exist is that A be non-singular. If A is symmetric, then so is A^{-1}. We also have

$$\begin{aligned} (A^{\mathrm{T}})^{-1} &= (A^{-1})^{\mathrm{T}} \\ \text{and} \quad (AB)^{-1} &= B^{-1}A^{-1} \end{aligned} \tag{1.6}$$

if A and B are square matrices of the same order.

A vector is a matrix with one row or one column and is printed in bold type. The notation $\mathbf{x}$ is assumed to denote a column vector, while a row vector is given as the transpose of a column vector.

A set of vectors $\mathbf{x}_1, \ldots, \mathbf{x}_p$ is said to be *linearly dependent* if there exist constants $c_1, \ldots, c_p$, which are not all zero, such that

$$\sum_{i=1}^{p} c_i\mathbf{x}_i = \mathbf{0}$$

Otherwise the vectors are said to be *linearly independent*. This definition leads on to the idea of a *rank* of a matrix, which is defined as the maximum number of rows which are linearly independent (or equivalently as the maximum number of columns which are independent). In other words, the rank is the dimension of the subspace spanned by vectors consisting of all the rows (or all the columns). If A is of order $(m \times n)$, then rank $(A) \leq \min(m, n)$. We find

$$\begin{aligned} \operatorname{rank}(A) &= \operatorname{rank}(A^{\mathrm{T}}) \\ &= \operatorname{rank}(AA^{\mathrm{T}}) \\ &= \operatorname{rank}(A^{\mathrm{T}}A) \end{aligned} \tag{1.7}$$

and

$$\operatorname{rank}(A) = \operatorname{rank}(BA) = \operatorname{rank}(AC) \tag{1.8}$$

where B, C are (square) non-singular matrices of appropriate order. If A is a (square) non-singular matrix of order $(p \times p)$, then A is of full rank p. But if A is singular, then the rows (and the columns) are linearly dependent and rank $(A) < p$.

Orthogonality. Two vectors $\mathbf{x}$, $\mathbf{y}$ of order $(p \times 1)$ are said to be *orthogonal* if

$$\mathbf{x}^{\mathrm{T}}\mathbf{y} = 0$$

They are further said to be *orthonormal* if

$$\mathbf{x}^{\mathrm{T}}\mathbf{x} = \mathbf{y}^{\mathrm{T}}\mathbf{y} = 1$$

A square matrix B is said to be *orthogonal* if

$$B^{\mathrm{T}}B = BB^{\mathrm{T}} = I \tag{1.9}$$

so that the rows of B are orthonormal. It is clear that B must be non-singular with

$$B^{-1} = B^{\mathrm{T}}$$

An orthogonal matrix may be interpreted as a linear transformation which consists of a rigid rotation or a rotation plus reflection, since it preserves distances and angles. The determinant of an orthogonal matrix must be ± 1. If the determinant is $(+1)$, the transformation is a rotation, while if the determinant is (-1), the transformation involves a reflection.

Quadratic forms. A *quadratic form* in p variables, $x_1, \ldots, x_p$ is a homogeneous function consisting of all possible second-order terms, namely

$$\begin{aligned} & a_{11}x_1^2 + \ldots + a_{pp}x_p^2 + a_{12}x_1x_2 + \ldots + a_{p-1,p}x_{p-1}x_p \\ & = \sum_{i,j} a_{ij}x_ix_j \end{aligned}$$

This can be conveniently written as $\mathbf{x}^{\mathrm{T}}A\mathbf{x}$, where $\mathbf{x}^{\mathrm{T}} = [x_1, \ldots, x_p]$ and a_{ij} is the (i, j)th element of A. The matrix A is usually taken to be symmetric.

A square matrix A and its associated quadratic form is called:

(*a*) *positive definite* if $\mathbf{x}^{\mathrm{T}}A\mathbf{x} > 0$ for every $\mathbf{x} \neq \mathbf{0}$;
(*b*) *positive semidefinite* if $\mathbf{x}^{\mathrm{T}}A\mathbf{x} \geq 0$ for every $\mathbf{x}$.

Positive-definite quadratic forms have matrices of full rank and we can then write

$$A = QQ^{\mathrm{T}} \tag{1.10}$$

where Q is non-singular. Then $\mathbf{y} = Q^{\mathrm{T}}\mathbf{x}$ transforms the quadratic form $\mathbf{x}^{\mathrm{T}}A\mathbf{x}$ to the reduced form $(y_1^2 + \ldots + y_p^2)$ which only involves squared terms.

If A is positive semidefinite of rank m $(< p)$, then A can also be expressed in the form of Equation (1.10), but with a matrix Q of order $(p \times m)$ which is of rank m. This is sometimes called the *Young–Householder factorization* of A, and will be used in classical scaling.

Eigenvalues and eigenvectors. Suppose the matrix Σ is of order $(p \times p)$. The eigenvalues of Σ (also called the *characteristic roots* or the *latent roots*) are the roots of the equation

$$|\Sigma - \lambda I| = 0 \tag{1.11}$$

which is a polynomial in λ of degree p. We denote the eigenvalues by $\lambda_1, \ldots, \lambda_p$. To each eigenvalue λ_i, there corresponds a vector $\mathbf{c}_i$, called an eigenvector, such that

$$\Sigma \mathbf{c}_i = \lambda_i \mathbf{c}_i \tag{1.12}$$

The eigenvectors are not unique as they contain an arbitrary scale factor, and so they are usually normalized so that $\mathbf{c}_i^{\mathrm{T}}\mathbf{c}_i = 1$. When there are equal eigenvalues, the corresponding eigenvectors can, and will, be chosen to be orthonormal. Some useful properties are as follows:

(*a*) $$\sum_{i=1}^{p} \lambda_i = \text{trace}\ (\Sigma) \tag{1.13}$$

(*b*) $$\prod_{i=1}^{p} \lambda_i = |\Sigma| \tag{1.14}$$

(*c*) If Σ is a real symmetric matrix, then its eigenvalues and eigenvectors are real.

(*d*) If, further, Σ is positive definite, then all the eigenvalues are strictly positive.

(*e*) If Σ is positive semidefinite of rank m $(< p)$, then Σ will have m positive eigenvalues and $(p - m)$ zero eigenvalues.

(*f*) For two unequal eigenvalues, the corresponding normalized eigenvectors are orthonormal.

(*g*) If we form a $(p \times p)$ matrix C, whose ith column is the normalized eigenvector $\mathbf{c}_i$, then $C^{\mathrm{T}}C = I$ and

$$C^{\mathrm{T}} \Sigma C = \Lambda \tag{1.15}$$

where Λ is a diagonal matrix whose diagonal elements are $\lambda_1, \ldots, \lambda_p$. This is called the *canonical reduction* of Σ. The matrix C transforms the quadratic form of Σ to the reduced form which only involves squared terms. Writing $\mathbf{x} = C\mathbf{y}$, we have

$$\begin{aligned} \mathbf{x}^{\mathrm{T}} \Sigma \mathbf{x} &= \mathbf{y}^{\mathrm{T}} C^{\mathrm{T}} \Sigma C \mathbf{y} \\ &= \mathbf{y}^{\mathrm{T}} \Lambda \mathbf{y} \\ &= \lambda_1 y_1^2 + \ldots + \lambda_m y_m^2 \end{aligned} \tag{1.16}$$

where $m = \text{rank}\ (\Sigma)$.

From Equation (1.15) we may also write

$$\Sigma = C\Lambda C^{\mathrm{T}} = \lambda_1 \mathbf{c}_1 \mathbf{c}_1^{\mathrm{T}} + \ldots + \lambda_m \mathbf{c}_m \mathbf{c}_m^{\mathrm{T}} \tag{1.17}$$

This is called the *spectral decomposition* of Σ.

Differentiation with respect to vectors. Suppose we have a differentiable function of p variables, say $f(x_1, \ldots, x_p)$. The notation $\partial f/\partial \mathbf{x}$ will be used to denote a column vector whose ith component is $\partial f/\partial x_i$.

Suppose that the function is the quadratic form $\mathbf{x}^{\mathrm{T}}\Sigma\,\mathbf{x}$, where Σ is a $(p \times p)$ symmetric matrix. Then it is left as an exercise for the reader to show that

$$\frac{\partial f}{\partial \mathbf{x}} = 2\Sigma\,\mathbf{x} \tag{1.18}$$

1.7 The general linear model

This book assumes that the reader has studied such topics as linear regression, multiple regression, simple experimental designs and the analysis of variance (ANOVA for short). Although these topics involve more than one variable, they are not usually covered in a course on multivariate analysis and so will not be discussed in detail in this book. All the topics are based on models which can be considered as special cases of what is called the *general linear model*. For completeness, we present the main results here.

The general linear model assumes that we have n uncorrelated observations such that

$$\begin{aligned} E(Y_r) &= \beta_0 + x_{r1}\beta_1 + \ldots + x_{rm}\beta_m \\ \mathrm{Var}(Y_r) &= \sigma^2 \end{aligned} \tag{1.19}$$

for $r = 1, 2, \ldots, n$. Here $\beta_0, \beta_1, \ldots, \beta_m$ and σ^2 are unknown parameters, while the 'coefficients' $\{x_{rj}\}$ may be one of two kinds. They may be actual observations on explanatory variables, so that x_{rj} denotes the rth observation on variable j. Then model (1.19) is the familiar multiple regression model with m explanatory variables. In particular, if $m = 1$, then model (1.19) reduces to a simple linear regression model with

$$E(Y_r) = \beta_0 + x_{r1}\beta_1 \tag{1.20}$$

It is more usual to write this in mean-corrected form as

$$E(Y_r) = \beta_0^* + (x_{r1} - \bar{x}_1)\beta_1 \quad \text{where } \bar{x}_1 = \sum_{r=1}^{n} x_{r1}/n \tag{1.20a}$$

When the model (1.19) is applied to data from experimental designs, the x's often denote indicator-type variables which only take the values zero or one, depending on which combination of factor levels is applied to get a particular observation on the dependent variable.

Model (1.19) has been further generalized (Nelder and Wedderburn, 1972) so as to cover the analysis of contingency tables and ANOVA with random effects.

Model (1.19) is often written in matrix form as

$$\mathbf{y} = \mathbf{1}\beta_0 + G\boldsymbol{\beta} + \mathbf{e} \tag{1.21}$$

where $\mathbf{y}^T = [y_1, \ldots, y_n]$, $\mathbf{1}$ denotes a $(n \times 1)$ vector of ones, $\boldsymbol{\beta}^T = [\beta_1, \ldots, \beta_m]$, and $\mathbf{e}^T = [e_1, \ldots, e_n]$ denotes the vector of 'errors'. The matrix G is of order $(n \times m)$ and its (r, j)th element is x_{rj} – or if the mean-corrected version of (1.19) is used, then the (r, j)th element of G is $(x_{rj} - \bar{x}_j)$. The matrix G is often called the *design matrix*.

The error terms are usually assumed to have mean zero, constant variance σ^2, and to be uncorrelated, although a more general version of the model does allow the errors to be correlated.

In multiple regression, the mean-corrected version of (1.19) is nearly always used, since the least-squares estimator of β_0 is then given simply by $\hat{\beta}_0 = \bar{y}$, while the least-squares estimator of $\boldsymbol{\beta}$ is obtained by solving the set of equations

$$G^T G \hat{\boldsymbol{\beta}} = G^T \mathbf{y} \tag{1.22}$$

which are often called the *normal equations.*

If $(G^T G)$ is of full rank, Equation (1.22) has a unique solution and may be readily solved to give

$$\hat{\boldsymbol{\beta}} = (G^T G)^{-1} G^T \mathbf{y} \tag{1.23}$$

and this turns out to be the best linear unbiased estimator (BLUE) for $\boldsymbol{\beta}$. The covariance matrix of $\hat{\boldsymbol{\beta}}$ (see Section 2.2) is then given by

$$\text{Var}(\hat{\boldsymbol{\beta}}) = \sigma^2 (G^T G)^{-1} \tag{1.24}$$

One practical problem arising in multiple regression is that of deciding how many of a set of possible explanatory variables are worth including in the model. Various algorithms are available. The simplest methods, called step-up or step-down methods, simply add or subtract one variable at a time, testing at each stage whether the change in fit is 'significant'. More complicated algorithms, such as that proposed by Beale, Kendall and Mann (1967), allow one to choose the 'best' subset of variables.

Finally, a warning about multiple regression. If one tries to include too many explanatory variables, there is always a danger that 'significant' relationships will be found which are actually spurious. This is particularly true when the explanatory variables cannot be controlled and are highly correlated. Although the matrix $(G^T G)$ is usually of full rank in multiple regression problems, it may be ill-conditioned if the explanatory variables are highly correlated, and this creates severe problems.

There are particular dangers associated with trying to fit multiple regression models to time-series data, as often seems to happen in econometrics. Here we find, not only that the explanatory variables are generally highly correlated, but also that the residuals are not uncorrelated. With a mis-specified model, one can easily derive 'nonsense' equations as clearly demonstrated by Granger and Newbold (1974).

Exercises

The following problems are intended to complement Section 1.6 by giving the reader some revision exercises in matrix algebra.

1.1 If $A = \begin{bmatrix} 1 & x & x^2 & 0 \\ 0 & 1 & x & x^2 \\ x^2 & 0 & 1 & x \\ x & x^2 & 0 & 1 \end{bmatrix}$

show that $|A| = 1 + x^4 + x^8$.

1.2 Find the inverse of the matrix $A = \begin{bmatrix} 2 & 0 & 0 & 0 \\ 0 & 4 & 0 & 0 \\ 0 & 0 & 3 & 1 \\ 0 & 0 & 1 & 5 \end{bmatrix}$

1.3 Find a (2×2) orthogonal matrix which can be interpreted as rotating the x–y axes clockwise through an angle θ.

1.4 Find an orthogonal (3×3) matrix whose first row is $(1/\sqrt{3})\ (1, 1, 1)$.

1.5 If A is a $(p \times p)$ matrix whose diagonal elements are all unity, and whose off-diagonal elements are all r, show that $|A| = (1 - r)^{p-1}\{1 + (p - 1)r\}$ and that A is non-singular except when $r = 1$ or $r = -1/(p - 1)$.
Show that the eigenvalues of A are such that $(p - 1)$ values are equal to $(1 - r)$ and the pth eigenvalue is $1 + (p - 1)r$.

1.6 Find the eigenvalues of $A = \begin{bmatrix} 2 & 1 & 1 \\ 1 & 2 & 1 \\ 1 & 1 & 2 \end{bmatrix}$

using the results of Exercise 1.5. Find one possible set of eigenvectors.

1.7 Find the symmetric matrix A corresponding to the quadratic form $\sum_{i=1}^{n} (x_i - \bar{x})^2$, and see if it is positive definite or positive semidefinite. Check that A is idempotent (i.e., $AA = A$) and that its eigenvalues are either zero or one. Show that A is singular and find its rank. (You should use the results of Exercise 1.5 again.)

1.8 Find the rank of the matrix $A = \begin{bmatrix} 1 & 2 & 3 & 4 \\ 3 & 6 & 9 & 12 \\ 4 & 3 & 2 & 1 \\ -1 & 3 & 7 & 11 \\ 8 & 6 & 4 & 2 \end{bmatrix}$

1.9 Find the spectral decomposition of the matrix

$$A = \begin{bmatrix} 1 & r \\ r & 1 \end{bmatrix}$$

CHAPTER TWO
Multivariate distributions

2.1 Multivariate, marginal and conditional distributions

The most basic concept in multivariate analysis is the idea of a multivariate probability distribution. We assume the reader is familiar with the definition of a (univariate) random variable and with standard probability distributions such as the normal distribution. This chapter extends these univariate ideas to the multivariate case. We discuss the general properties of multivariate distributions and consider some particular examples including the multivariate normal distribution.

We will denote a p-dimensional random variable, sometimes called a random vector, by $\mathbf{X}$ where

$$\mathbf{X}^{\mathrm{T}} = [X_1, \ldots, X_p]$$

and $X_1, \ldots, X_p$ are univariate random variables. We remind the reader that vectors are printed in bold type, but matrices are in ordinary type. Thus the notation X, which was introduced in Chapter 1 to denote the $(n \times p)$ data matrix of observed values, should not be confused with a vector random variable, $\mathbf{X}$. The data matrix might possibly be confused with a univariate random variable, but it should always be clear from the context as to which meaning is intended.

We shall be mainly concerned with the case where the variables are *continuous*, but it is probably easier to understand the ideas if we start by considering the case where the components of $\mathbf{X}$ are *discrete* random variables. In this case, the (joint) distribution of $\mathbf{X}$ is described by the joint probability function $P(x_1, \ldots x_p)$, where

$$P(x_1, \ldots, x_p) = \mathrm{Prob}(X_1 = x_1, \ldots, X_p = x_p)$$

To save space, we write $P(x_1, \ldots, x_p)$ as $P(\mathbf{x})$. Note that lower-case letters indicate particular values of the corresponding random variables.

The function $P(\mathbf{x})$ must satisfy two conditions similar to those required in the univariate case, namely that

$$P(\mathbf{x}) \geq 0 \qquad \text{for every } \mathbf{x}$$

and

$$\sum P(\mathbf{x}) = 1$$

where the summation is over all possible values of $\mathbf{x}$.

From the joint distribution, it is often useful to calculate two other types of distribution, namely *marginal* distributions and *conditional* distributions.

Suppose we are interested in the distribution of one component of $\mathbf{X}$, say X_i, without regard to the values of the other variables. Then the probability distribution of X_i can be obtained from the joint distribution by summing over all the other variables. Thus

$$\text{Prob}(X_i = x_i) = \sum P(x_1, \ldots, x_i, \ldots, x_p)$$

where the summation is over all $\mathbf{x}$ such that the ith component is fixed to be x_i, in other words over $x_1, \ldots, x_{i-1}, x_{i+1}, \ldots, x_p$.

When the distribution of a single variable is obtained from a joint distribution by summing over all other variables, then it is usually called a *marginal* distribution. If the joint distribution is equal to the product of all the marginal distributions for every $\mathbf{x}$, so that

$$P(\mathbf{x}) = \prod_{i=1}^{p} P_i(x_i)$$

where $P_i(x_i)$ denotes the marginal distribution of X_i, then the random variables are said to be *independent*.

We also note that marginal joint distributions are possible by summing over less than $(p-1)$ variables. For example, the marginal joint distribution of X_1 and X_2 is obtained by summing the joint distribution over all variables from X_3 to X_p.

If some of the variables are set equal to specified constant values, then the distribution of the remaining variables is called a *conditional* distribution.

For a conditional probability, recall that for two events, A and B,

$$\begin{aligned}\text{Prob}(\text{A}|\text{B}) &= \text{probability of event A given that B has occurred} \\ &= P(\text{A} \cap \text{B})/P(\text{B}),\end{aligned}$$

where $\text{A} \cap \text{B}$ denotes the intersection of events A and B, meaning that both events occur. By analogy, we find that the conditional distribution of a random variable is given by the ratio of the joint distribution to the appropriate marginal distribution.

In the case $p = 2$, the conditional distribution of X_1, given that X_2 takes the particular value x_2, is given by

$$\begin{aligned}P(x_1|x_2) &= \text{Prob}(X_1 = x_1|X_2 = x_2) \\ &= P(x_1, x_2)/P_2(x_2)\end{aligned}$$

where $P_2(x_2)$ is the marginal distribution of X_2. More generally,

$$P(x_1, \ldots, x_k|x_{k+1}, \ldots, x_p) = P(\mathbf{x})/P_{\text{M}}(x_{k+1}, \ldots, x_p)$$

where $P_{\text{M}}(x_{k+1}, \ldots, x_p)$ denotes the marginal joint distribution of $X_{k+1}, \ldots, X_p$.

Example 2.1. A coin is tossed four times. Let

$$X_1 = \text{number of heads in first 2 throws}$$
$$X_2 = \text{number of heads in last 3 throws.}$$

Find the joint distribution of X_1 and X_2, the marginal distribution of X_1, and the conditional distribution of X_1 given that $X_2 = 2$.

Using elementary probability, we find that the joint distribution of X_1 and X_2 is as shown in Table 2.1.

Table 2.1 The joint distribution

X_2	X_1			Marginal distribution of X_2
	0	1	2	
0	$\frac{1}{16}$	$\frac{1}{16}$	0	$\frac{1}{8}$
1	$\frac{1}{8}$	$\frac{3}{16}$	$\frac{1}{16}$	$\frac{3}{8}$
2	$\frac{1}{16}$	$\frac{3}{16}$	$\frac{1}{8}$	$\frac{3}{8}$
3	0	$\frac{1}{16}$	$\frac{1}{16}$	$\frac{1}{8}$
Marginal distribution of X_1	$\frac{1}{4}$	$\frac{1}{2}$	$\frac{1}{4}$	

Note that the sum of the joint probabilities is 1. The marginal distributions of X_1 and X_2 can be found by summing the column and row joint probabilities respectively. Of course, in this simple case we could write down the (marginal) distribution of X_1 (and of X_2) directly. The conditional distribution of X_1, given that $X_2 = 2$, is obtained by looking at the row of joint probabilities where $X_2 = 2$ and normalizing them so that they sum to 1. This is done by dividing by the row total, which is the appropriate marginal probability. Thus

$$\text{Prob}(X_1 = 0 | X_2 = 2) = \frac{1/16}{3/8} = 1/6$$
$$\text{Prob}(X_1 = 1 | X_2 = 2) = 1/2$$
$$\text{Prob}(X_1 = 2 | X_2 = 2) = 1/3$$

Note that these conditional probabilities sum to unity. □

Continuous variables. In the univariate case, the distribution of a continuous random variable may be described by the cumulative distribution function (abbreviated c.d.f.) or by its derivative called the probability density function (abbreviated p.d.f.). For continuous multivariate distributions, we may define suitable multivariate analogues of these functions. The joint c.d.f., which we will denote by $F(x_1, \ldots, x_p)$, is defined by

$$F(x_1, \ldots, x_p) = \text{Prob}(X_1 \leq x_1, \ldots, X_p \leq x_p).$$

The joint p.d.f., which we denote by $f(x_1, \ldots, x_p)$ or $f(\mathbf{x})$, is then given by the pth partial derivative

$$f(\mathbf{x}) = \frac{\partial^p F(x_1, \ldots, x_p)}{\partial x_1 \partial x_2 \ldots \partial x_p}$$

assuming that $F(\mathbf{x})$ is absolutely continuous.

The joint p.d.f. satisfies the restrictions:

(*a*) $f(\mathbf{x}) \geq 0$ for every $\mathbf{x}$;

(*b*) $\int_{-\infty}^{\infty} \ldots \int_{-\infty}^{\infty} f(\mathbf{x}) \mathrm{d}x_1 \ldots \mathrm{d}x_p = 1.$

As in the univariate case, the joint p.d.f. is not a probability, but rather probabilities can be found by integrating over the required subset of p-space. The reader who is unfamiliar with the joint p.d.f. should relate the above definitions to the bivariate case where $p = 2$. Here one can think of the two variables as being defined along perpendicular axes in a plane. For any point (x_1, x_2) in this plane, the joint p.d.f. gives the height of a surface above this plane. The total volume under this surface (or hill) is defined to be 1 and is the total probability. For any area in the plane, the corresponding volume underneath the surface gives the probability of getting a bivariate observation in the given area.

Marginal and conditional distributions may easily be defined in the continuous case. The marginal p.d.f. of one component of $\mathbf{X}$, say X_i, may be found from the joint p.d.f. by integrating out all the other variables. Thus

$$f_i(x_i) = \int_{-\infty}^{\infty} \ldots \int_{-\infty}^{\infty} f(\mathbf{x}) \mathrm{d}x_1 \ldots \mathrm{d}x_{i-1} \mathrm{d}x_{i+1} \ldots \mathrm{d}x_p$$

The random variables are independent if the joint p.d.f. is equal to the product of all the marginal p.d.f.s for every $\mathbf{x}$. We also note that marginal joint distributions are possible by integrating out less than $(p-1)$ variables.

The density functions of conditional continuous distributions can be found by dividing the joint p.d.f. by the appropriate marginal p.d.f. This is clearly analogous to the discrete case. Thus, in the case $p = 2$, the conditional p.d.f. of X_1, given that X_2 takes the particular value x_2, will be denoted by $h(x_1|x_2)$ and is given by

$$h(x_1|x_2) = f(x_1, x_2)/f_2(x_2) \tag{2.1}$$

More generally, the conditional joint p.d.f. of $X_1, \ldots, X_k$ given $X_{k+1} = x_{k+1}, \ldots, X_p = x_p$ is given by

$$h(x_1, \ldots, x_k | x_{k+1}, \ldots, x_p) = f(\mathbf{x})/f_{\mathrm{M}}(x_{k+1}, \ldots, x_p)$$

where $f_{\mathrm{M}}(x_{k+1}, \ldots, x_p)$ denotes the marginal joint p.d.f. of $X_{k+1}, \ldots, X_p$.

Example 2.2. To illustrate the definitions in the continuous case, consider the bivariate distribution with joint p.d.f.

$$f(x_1, x_2) = \begin{cases} 2 & 0 < x_1 < x_2 < 1 \\ 0 & \text{otherwise} \end{cases}$$

Thus the density function is constant over the triangle illustrated in Fig. 2.1. What are the marginal distributions of X_1 and X_2, and are the random variables independent?

The random variable X_1 is defined over the range (0, 1), where its density function is given by

$$f_1(x_1) = \int_{x_1}^{1} 2\mathrm{d}x_2$$

Thus

$$f_1(x_1) = \begin{cases} 2(1 - x_1) & 0 < x_1 < 1 \\ 0 & \text{otherwise} \end{cases}$$

Similarly, for X_2 we find

$$f_2(x_2) = \begin{cases} 2x_2 & 0 < x_2 < 1 \\ 0 & \text{otherwise} \end{cases}$$

The two random variables are not independent since

$$f(x_1, x_2) \neq f_1(x_1) f_2(x_2)$$

The fact that the random variables are not independent is obvious since if we observe the value $X_2 = 1/2$, say, then X_1 is constrained to the range $(0, \frac{1}{2})$ rather than the whole range (0, 1).

We now derive one conditional distribution as an exercise. In finding conditional distributions, it is important to start by finding the correct range of possible values as this is not always obvious. Suppose in our example we want to find the conditional distribution of X_1 given that X_2 takes the value 3/4.

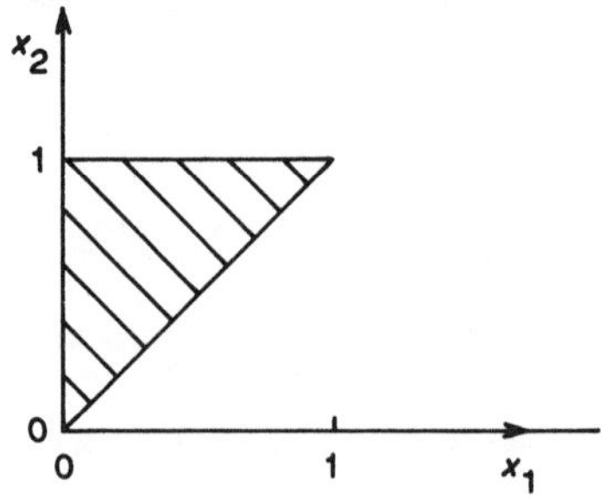

Figure 2.1

Looking at Fig. 2.1, we see that the conditional distribution must be defined over the range (0, 3/4). Using Equation (2.1), we find that the conditional density function must be a constant over this range so that

$$h(x_1|X_2 = 3/4) = \begin{cases} 4/3 & 0 < x_1 < 3/4 \\ 0 & \text{otherwise} \end{cases}$$

Note that this integrates to unity over the range (0, 3/4). □

2.2 Means, variances, covariances and correlations

In the univariate case, it is often useful to summarize a probability distribution by giving the first two moments, namely the mean and variance (or its square root, the standard deviation). To summarize multivariate distributions, we need to find the mean and variance of each of the p variables, together with a measure of the way each pair of variables is related. The latter target is achieved by calculating a set of quantities called *covariances*, or their standardized counterparts called *correlations*. These quantities will now be defined.

Means. The mean vector $\boldsymbol{\mu}^{\mathrm{T}} = [\mu_1, \ldots, \mu_p]$ is such that

$$\mu_i = E(X_i) = \int_{-\infty}^{\infty} x f_i(x)\mathrm{d}x \tag{2.2}$$

is the mean of the ith component of $\mathbf{X}$. This definition is given for the case where X_i is continuous. If X_i is discrete, then $E(X_i)$ is given by $\sum x P_i(x)$, where $P_i(x)$ is the (marginal) probability distribution of X_i.

Variances. The variance of the ith component of X is given by

$$\begin{aligned} \mathrm{Var}(X_i) &= E[(X_i - \mu_i)^2] \\ &= E(X_i^2) - \mu_i^2 \end{aligned} \tag{2.3}$$

This is usually denoted by σ_i^2 in the univariate case, but in order to tie in with the covariance notation given below, we will usually denote it by σ_{ii} in the multivariate case.

Covariances. The covariance of two variables X_i and X_j is defined by

$$\mathrm{Cov}(X_i, X_j) = E[(X_i - \mu_i)(X_j - \mu_j)] \tag{2.4}$$

Thus, it is the product moment of the two variables about their respective means. In particular, if $i = j$, we note that the covariance of a variable with itself is simply the variance of the variable. Thus, there is really no need to define variance separately in the multivariate case, as it is a special case of covariance.

The covariance of X_i and X_j is usually denoted by σ_{ij}. Then, if $i = j$, the variance of X_i is denoted by σ_{ii}, as noted above.

Equation (2.4) is often written in the equivalent alternative form

$$\sigma_{ij} = E[X_i X_j] - \mu_i \mu_j \tag{2.5}$$

The covariance matrix. With p variables, there are p variances and $\frac{1}{2}p(p-1)$ covariances, and these quantities are all second moments. It is often useful to present these quantities in a $(p \times p)$ matrix, denoted by Σ, whose (i, j)th element is σ_{ij}. Thus,

$$\Sigma = \begin{bmatrix} \sigma_{11} & \sigma_{12} & \cdots & \sigma_{1p} \\ \sigma_{21} & \sigma_{22} & \cdots & \sigma_{2p} \\ \vdots & & & \\ \sigma_{p1} & \sigma_{p2} & \cdots & \sigma_{pp} \end{bmatrix}$$

This matrix is variously called the *dispersion* matrix, the *variance–covariance* matrix, or simply the *covariance* matrix, and we will use the latter term. The diagonal terms of Σ are the variances, while the off-diagonal terms, the covariances, are such that $\sigma_{ij} = \sigma_{ji}$. Thus the matrix is symmetric.

Using Equations (2.4) and (2.5), we can express Σ in two alternative useful forms, namely

$$\begin{aligned} \Sigma &= E[(\mathbf{X} - \boldsymbol{\mu})(\mathbf{X} - \boldsymbol{\mu})^{\mathrm{T}}] \\ &= E[\mathbf{X}\mathbf{X}^{\mathrm{T}}] - \boldsymbol{\mu}\boldsymbol{\mu}^{\mathrm{T}} \end{aligned} \tag{2.6}$$

Linear compounds. Perhaps the main use of covariances is as a stepping stone to the calculations of correlations (see below), but they are also useful for a variety of other purposes. Here we illustrate their use in finding the variance of any linear combination of the components of **X**. Such combinations arise in a variety of situations, and in Chapter 6 we will define them as being 'linear compounds', reserving the term 'linear combination' for a weighted sum of vectors rather than scalars. Consider the general linear compound

$$Y = \mathbf{a}^{\mathrm{T}}\mathbf{X}$$

where $\mathbf{a}^{\mathrm{T}} = [a_1, \ldots, a_p]$ is a vector of constants. Then Y is a univariate random variable. Its mean is clearly given by

$$E(Y) = \mathbf{a}^{\mathrm{T}}\boldsymbol{\mu} \tag{2.7}$$

while its variance is given by

$$\mathrm{Var}(Y) = E[\{\mathbf{a}^{\mathrm{T}}(\mathbf{X} - \boldsymbol{\mu})\}^2]$$

As $\mathbf{a}^{\mathrm{T}}(\mathbf{X} - \boldsymbol{\mu})$ is a scalar and therefore equal to its transpose, we can express $\mathrm{Var}(Y)$ in terms of Σ, using Equation (2.6), as

$$\begin{aligned} \mathrm{Var}(Y) &= E[\mathbf{a}^{\mathrm{T}}(\mathbf{X} - \boldsymbol{\mu})(\mathbf{X} - \boldsymbol{\mu})^{\mathrm{T}}\mathbf{a}] \\ &= \mathbf{a}^{\mathrm{T}} E[(\mathbf{X} - \boldsymbol{\mu})(\mathbf{X} - \boldsymbol{\mu})^{\mathrm{T}}]\mathbf{a} \\ &= \mathbf{a}^{\mathrm{T}}\Sigma\,\mathbf{a} \end{aligned} \tag{2.8}$$

To illustrate the last formula, consider the case $p = 2$, where $\mathbf{a}^T\mathbf{X} = a_1X_1 + a_2X_2$. The reader will probably already know that

$$\begin{aligned}\mathrm{Var}(a_1X_1 + a_2X_2) &= \mathrm{Var}(a_1X_1) + \mathrm{Var}(a_2X_2) + 2\mathrm{Cov}(a_1X_1, a_2X_2)\\ &= a_1^2\mathrm{Var}(X_1) + a_2^2\mathrm{Var}(X_2) + 2a_1a_2\mathrm{Cov}(X_1, X_2)\end{aligned}$$

This may be written in the form

$$[a_1 \quad a_2]\begin{bmatrix}\mathrm{Var}(X_1) & \mathrm{Cov}(X_1, X_2)\\ \mathrm{Cov}(X_1, X_2) & \mathrm{Var}(X_2)\end{bmatrix}\begin{bmatrix}a_1\\ a_2\end{bmatrix}$$

Equations (2.7) and (2.8) may be generalized by noting that if A is any $(p \times m)$ matrix of constants, then it can easily be shown that $A^T\mathbf{X}$ (which is a $(m \times 1)$ vector random variable) has a mean vector and covariance matrix given by

$$\begin{aligned}E(A^T\mathbf{X}) &= A^T\boldsymbol{\mu}\\ \mathrm{Var}(A^T\mathbf{X}) &= A^T\Sigma A\end{aligned} \tag{2.9}$$

Correlations. Although covariances are useful for many mathematical purposes, they are rarely used as descriptive statistics. If two variables are related in a linear way, then the covariance will be positive or negative depending on whether the relationship has a positive or negative slope. But the size of the coefficient is difficult to interpret because it depends on the units in which the two variables are measured. Thus the covariance is often standardized by dividing by the product of the standard deviations of the two variables to give a quantity called the *correlation coefficient*. The correlation between variables X_i and X_j will be denoted by ρ_{ij}, and is given by

$$\rho_{ij} = \sigma_{ij}/\sigma_i\sigma_j \tag{2.10}$$

where σ_i denotes the standard deviation of X_i.

It can be shown that ρ_{ij} must lie between -1 and $+1$, by using the fact that $\mathrm{Var}(aX_i + bX_j) \geq 0$ for every a, b, and putting $a = \sqrt{\mathrm{Var}(X_j)}$ and $b = \pm\sqrt{\mathrm{Var}(X_i)}$.

The correlation coefficient provides a measure of the linear association between two variables. The coefficient is positive if the relationship between the two variables has a positive slope so that 'high' values of one variable tend to go with 'high' values of the other variable. Conversely, the coefficient is negative if the relationship has a negative slope.

If two variables are independent than their covariance, and hence their correlation, will be zero. But it is important to note that the converse of this statement is not true. It is possible to construct examples where two variables have zero correlation and yet are dependent on one another, often in a non-linear way (see Exercise 2.3). This emphasizes the fact that the correlation coefficient is of no use as a descriptive statistic (and may be positively misleading) if the relationship between two variables is of a non-linear form. However, if the two variables follow a bivariate normal distribution as described in Section 2.4, then it turns out that zero correlation *does* imply independence.

The correlation matrix. With p variables, there are $p(p-1)/2$ distinct correlations. It is often useful to present them in a $(p \times p)$ matrix whose (i, j)th element is defined to be ρ_{ij}. This matrix, called the correlation matrix, will be denoted by P which is the Greek letter for capital rho. The diagonal terms of P are unity, and the off-diagonal terms are such that P is symmetric.

In order to relate the covariance and correlation matrices, let us define a $(p \times p)$ diagonal matrix D, whose diagonal terms are the standard deviations of the components of $\mathbf{X}$, so that

$$D = \begin{bmatrix} \sigma_1 & 0 & \dots & 0 \\ 0 & \sigma_2 & \dots & 0 \\ \vdots & & & \\ 0 & 0 & \dots & \sigma_p \end{bmatrix}$$

Then the covariance and correlation matrices are related by

$$\Sigma = DPD$$

or

$$\mathrm{P} = D^{-1} \Sigma D^{-1} \tag{2.11}$$

where the diagonal terms of the matrix D^{-1} are the reciprocals of the respective standard deviations.

The rank of Σ *and* P. We complete this section with a slightly more advanced discussion of the matrix properties of Σ and P, and in particular of their rank.

Firstly, we show that both Σ and P are positive semidefinite. As any variance must be non-negative, we have that

$$\mathrm{Var}(\mathbf{a}^T\mathbf{X}) \geq 0 \qquad \text{for every } \mathbf{a}$$

But $\mathrm{Var}(\mathbf{a}^T\mathbf{X}) = \mathbf{a}^T \Sigma \mathbf{a}$, and so Σ must be positive semidefinite. We also note that Σ is related to P by Equation (2.11), where D is non-singular, and so it follows (see Exercise 2.6) that P is also positive semidefinite.

Because D is non-singular, we may also use Equations (2.11) and (1.8) to show that the rank of P is the same as the rank of Σ. This rank must be less than or equal to p.

If Σ (and hence P) is of full rank p, then Σ (and hence P) will be positive definite, as in this case, $\mathrm{Var}(\mathbf{a}^T\mathbf{X}) = \mathbf{a}^T \Sigma \mathbf{a}$ is strictly greater than zero for every $\mathbf{a} \neq \mathbf{0}$. But if $\mathrm{rank}(\Sigma) < p$, then Σ (and hence P) will be singular, and this indicates a linear constraint on the components of $\mathbf{X}$. This means that there exists a vector $\mathbf{a} \neq \mathbf{0}$ such that $\mathbf{a}^T\mathbf{X}$ is identically equal to a constant. Then $\mathrm{Var}(\mathbf{a}^T\mathbf{X}) = \mathbf{a}^T \Sigma \mathbf{a}$ is zero, indicating that Σ is positive semidefinite rather than positive definite.

To illustrate this last point, suppose that $p = 3$ and there is a linear constraint on the three variables such that $X_1 = X_2 + X_3$. Then $\mathrm{Var}(X_1 - X_2 - X_3)$ must be zero, where in this case $\mathbf{a}^T = [1, -1, -1]$. Obviously, one of the three variables is effectively redundant so that the dimensionality is 2 rather than 3. This is reflected in the rank of Σ (or of P) which will also be 2. It turns out that the rank of Σ is generally a useful guide to the effective dimensionality of the situation, as $[p - \text{rank}(\Sigma)]$ is equal to the number of independent linear constraints on the components of $\mathbf{X}$.

When $\text{rank}(\Sigma) < p$, the components of $\mathbf{X}$ are sometimes said to be 'linearly dependent', using this term in its algebraic sense. However, statisticians often use this term to mean a linear relationship between the *expected values* of the random variables. It needs to be emphasized that a constraint of the latter type will generally *not* produce a singular Σ. If two variables are correlated, it does not mean that one of them is redundant, although if the correlation is very high then one of them may be 'nearly redundant' and the covariance matrix will be 'nearly singular'.

Example 2.2 (*continued*). To find the moments of the distribution considered earlier in Example 2.2, we note that the centre of gravity of a uniform triangle as in Fig. 2.1 will be at the point (1/3, 2/3). Thus by elementary geometry we must have $E(X_1) = 1/3$ and $E(X_2) = 2/3$. This can easily be confirmed using Equation (2.2). For example,

$$E(X_1) = \int_0^1 x_1 f_1(x_1) = \int_0^1 x_1 2(1 - x_1)\mathrm{d}x_1 = 1/3.$$

The covariance of X_1 and X_2 is given by Equation (2.5), namely

$$\begin{aligned}\sigma_{12} &= E(X_1X_2) - E(X_1)E(X_2)\\ &= \int_0^1 \left[\int_0^{x_2} x_1x_2 2\mathrm{d}x_1\right]\mathrm{d}x_2 - 2/9\\ &= 1/36 \qquad \text{after some algebra.}\end{aligned}$$

As the variances of X_1 and X_2 are both found from Equation (2.3) to be 1/18, the correlation of X_1 and X_2 is given by $18/36 = 1/2$, using Equation (2.10). Intuitively we expect the correlation to be positive since, looking at Fig. 2.1, 'high' values of X_1 tend to go with 'high' values of X_2. □

Example 2.3. Suppose that the first $(p - 1)$ components of a p-dimensional random variable $\mathbf{X}$ are independent random variables, $X_1, X_2, \ldots, X_{p-1}$, each with the same variance σ^2, and that the pth component of $\mathbf{X}$ is given by $X_p = \sum_{i=1}^{p-1} X_i$. Find the covariance and correlation matrices for $\mathbf{X}$ and show that both matrices are singular.

The only non-zero covariances are between X_p and the other $(p - 1)$ variables. For $i = 1, \ldots, p - 1$, we have

$$\begin{aligned}\mathrm{Cov}(X_i, X_p) &= \mathrm{Cov}(X_i, X_i)\\ &= \mathrm{Var}(X_i)\\ &= \sigma^2\end{aligned}$$

Now

$$\begin{aligned}\operatorname{Var}(X_p) &= \operatorname{Var}\sum_{i=1}^{p-1} X_i \\ &= \sum_{i=1}^{p-1}\operatorname{Var}(X_i) \qquad \text{as } \{X_i\} \text{ are independent} \\ &= (p-1)\sigma^2\end{aligned}$$

Thus

$$\Sigma = \begin{bmatrix} \sigma^2 & 0 & \dots & 0 & \sigma^2 \\ 0 & \sigma^2 & \dots & 0 & \sigma^2 \\ \vdots & & & & \\ \sigma^2 & \sigma^2 & \dots & \sigma^2 & (p-1)\sigma^2 \end{bmatrix}$$

and

$$P = \begin{bmatrix} 1 & 0 & \dots & 0 & 1/\sqrt{(p-1)} \\ 0 & 1 & \dots & 0 & 1/\sqrt{(p-1)} \\ \vdots & & & & \\ 1/\sqrt{(p-1)} & 1/\sqrt{(p-1)} & \dots & & 1 \end{bmatrix}$$

It is easy to demonstrate that Σ is singular by showing that the determinant of Σ is zero. This can be done by subtracting the sum of the first $(p-1)$ columns from the pth column. The fact that Σ is singular and hence not of full rank p is in any case obvious from the fact that X_p is linearly related to the other random variables, so that one of the variables is effectively redundant. The rank of Σ is in fact $(p-1)$. It then follows that P must also be singular of rank $(p-1)$.

2.3 The multivariate normal distribution

The most commonly used multivariate distribution is the one called the multivariate normal distribution, which we abbreviate by MVN distribution. This section defines the distribution, while Chapter 6 discusses its properties in much more detail.

First we recall that a univariate normal random variable X, with mean μ and variance σ^2, has density function

$$f(x) = \frac{1}{\sqrt{(2\pi)}\,\sigma}\exp[-(x-\mu)^2/2\sigma^2] \tag{2.12}$$

and we write $X \sim \mathrm{N}(\mu, \sigma^2)$.

In the multivariate case, we say that a p-dimensional random variable $\mathbf{X}$ follows the multivariate normal distribution if its joint p.d.f. is of the form

$$f(\mathbf{x}) = \frac{1}{(2\pi)^{p/2}|\Sigma|^{1/2}}\exp[-\tfrac{1}{2}(\mathbf{x}-\boldsymbol{\mu})^{\mathrm{T}}\Sigma^{-1}(\mathbf{x}-\boldsymbol{\mu})] \tag{2.13}$$

where Σ is any $(p \times p)$ symmetric positive definite matrix.

It is not immediately obvious that Equation (2.13) is the natural generalization of Equation (2.12). However, we note that Equation (2.13) reduces to Equation (2.12) in the case $p = 1$. Moreover, if $X_1, \ldots, X_p$ are independent random variables where $X_i \sim N(\mu_i, \sigma_i^2)$, then their joint p.d.f. is simply the product of the appropriate (marginal) density functions, so that

$$f(x_1, \ldots, x_p) = \frac{1}{(2\pi)^{p/2} \prod_{i=1}^{p} \sigma_i} \exp\left[-\tfrac{1}{2} \sum_{i=1}^{p} \left(\frac{x_i - \mu_i}{\sigma_i} \right)^2 \right] \qquad (2.14)$$

In this case $\mathbf{X}^T = [X_1, \ldots, X_p]$ has mean $\boldsymbol{\mu}^T = [\mu_1, \ldots, \mu_p]$ and covariance matrix

$$\Sigma = \begin{bmatrix} \sigma_1^2 & 0 & \ldots & 0 \\ 0 & \sigma_2^2 & \ldots & 0 \\ \vdots & & & \\ 0 & 0 & \ldots & \sigma_p^2 \end{bmatrix}$$

and we find that Equation (2.14) can be rewritten in the form of Equation (2.13). Thus, the case of independent normal variables is a special case of the formula given in Equation (2.13). But of course the components of $\mathbf{X}$ do not generally need to be independent and so Σ does not have to be diagonal, provided that it is symmetric and positive definite. The requirement that Σ be positive definite can be thought of as the multivariate equivalent of the condition that $\sigma^2 > 0$ in Equation (2.12).

Thus far we have not shown that Equation (2.13) defines a proper distribution. Now it is clear that $f(\mathbf{x}) \geq 0$ for every $\mathbf{x}$, and it is also straightforward, though algebraically tedious, to check that $\int_{\mathbf{x}} f(\mathbf{x}) dx_1 \ldots dx_p = 1$ for every $\boldsymbol{\mu}$ and for every Σ which is symmetric and positive definite. After some algebra, it is also possible to show that $E(\mathbf{X}) = \boldsymbol{\mu}$ and that Σ is the covariance matrix for $\mathbf{X}$. Thus the parameters in $\boldsymbol{\mu}$ and Σ have an immediate interpretation, and we write $\mathbf{X} \sim N_p(\boldsymbol{\mu}, \Sigma)$, where p denotes the dimension of $\mathbf{X}$, $\boldsymbol{\mu}$ denotes the mean vector and Σ denotes the covariance matrix.

The definition of the MVN distribution via Equation (2.13) requires the covariance matrix to be non-singular so that Σ^{-1} exists. This excludes cases where there is a linear constraint on the $\{X_i\}$, though it does allow linear relationships between the expected values of $\{X_i\}$. A more general definition due to C. R. Rao is given in Chapter 6 and this does not exclude linear constraints on the $\{X_i\}$.

The importance of the MVN distribution arises partly from the multivariate form of the central limit theorem, and partly from the fact that some multivariate data can be approximated by the MVN distribution. Further discussion is deferred to Chapter 6.

2.4 The bivariate normal distribution

An important special case of the multivariate normal distribution arises when there are just two variables. In this case we denote the mean vector by $\boldsymbol{\mu}^{\mathrm{T}} = [\mu_1, \mu_2]$, while the covariance matrix is usually expressed in the form

$$\Sigma = \begin{bmatrix} \sigma_1^2 & \rho\sigma_1\sigma_2 \\ \rho\sigma_1\sigma_2 & \sigma_2^2 \end{bmatrix} \tag{2.15}$$

where ρ denotes the correlation between the two variables.

If the reader works out Σ^{-1} and $|\Sigma|^{1/2}$, he will find that the joint p.d.f. given by Equation (2.13) can be expressed in the form

$$f(x_1, x_2) = \frac{1}{2\pi\sigma_1\sigma_2\sqrt{(1-\rho^2)}} \exp\left\{ -\frac{1}{2(1-\rho^2)}\left[\left(\frac{x_1-\mu_1}{\sigma_1}\right)^2 - 2\rho\left(\frac{x_1-\mu_1}{\sigma_1}\right)\left(\frac{x_2-\mu_2}{\sigma_2}\right) + \left(\frac{x_2-\mu_2}{\sigma_2}\right)^2\right]\right\} \tag{2.16}$$

This rather complicated looking expression depends on five parameters, the two means, the two variances and the correlation. Looking at Equation (2.15), we see that Σ is non-singular, and hence positive definite, provided that $|\rho| < 1$. If $\rho = +1$ or -1, the two variables are linearly related and the observations are effectively one-dimensional, so that a degenerate form of the bivariate normal distribution arises. If $\rho = 0$, then we see that Equation (2.16) reduces to the product of two univariate normal density functions and the two variables are in fact independent. As we have already remarked, zero correlation does imply independence for the multivariate normal distribution, though it may not do so for other multivariate distributions.

To understand Equation (2.16), note that the variables x_1 and x_2 only appear in the exponential expression. Thus $f(x_1, x_2)$ is constant when

$$z_1^2 - 2\rho z_1 z_2 + z_2^2 = \text{constant}$$

where $z_i = (x_i - \mu_i)/\sigma_i$. For $|\rho| < 1$, this is the equation of an *ellipse*. When $\rho > 0$, the major axes of the ellipses have a positive slope, as illustrated in Fig. 2.2, while if $\rho < 0$, the major axes have a negative slope. The joint p.d.f. is clearly a maximum at $\boldsymbol{\mu}$.

The reader will probably already have met the bivariate normal distribution when studying linear regression. For this distribution it can be shown that the

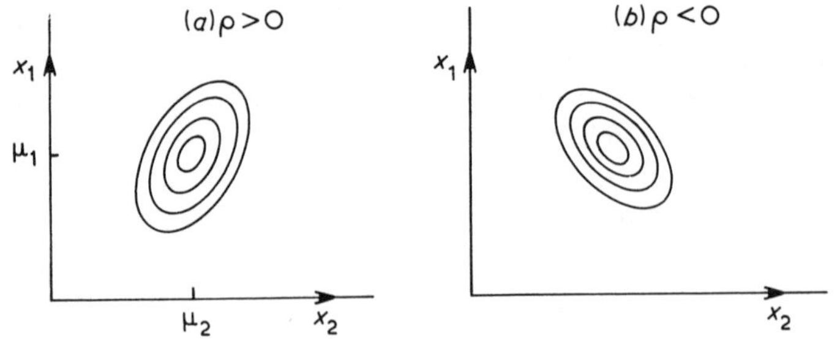

Figure 2.2. Ellipses on which the joint p.d.f. is constant for (*a*) $\rho > 0$, (*b*) $\rho < 0$.

regression equations of X_1 on X_2 and of X_2 on X_1 are both straight lines. It can also be shown that the marginal distributions of both X_1 and X_2 are normal and that all conditional distributions are normal. These results will be proved more generally in Chapter 6.

2.5 Other multivariate distributions

A variety of different classes of multivariate distributions are available in the literature apart from the multivariate normal. Some of these are used very rarely, but it is useful to know of their existence. We first consider discrete distributions and then continuous ones.

2.5.1 *Multivariate discrete distributions*

One of the best sources of information on this topic is Johnson and Kotz (1969, Chapter 11). These authors consider a variety of distributions including the multinomial distribution, the negative multinomial distribution, various multivariate Poisson distributions, the multivariate hypergeometric distribution, the multivariate logarithmic series distribution, the bivariate Neyman Type A distribution, and so on. Although we will not describe them here, this list does at least indicate the rich variety of distributions available.

As just one example, let us define the well-known *multinomial* distribution. Consider a series of n independent trials, in each of which just one of k mutually exclusive and exhaustive events $E_1, \ldots, E_k$, must be observed. Assume that the probability of observing E_i is given by p_i, where $\sum_{i=1}^{k} p_i = 1$, and that n_i occurrences of event E_i are observed, where $\sum_{i=1}^{k} n_i = n$. Then it can be shown that the joint distribution of $n_1, \ldots, n_k$ is given by

$$P(n_1, \ldots, n_k) = n! \prod_{j=1}^{k} p_j^{n_j}/n_j!$$

Although the multinomial distribution is technically a multivariate distribution, it is only concerned with one factor and so it is not generally used in multivariate analysis. Discrete multivariate data arise when one is interested in several discrete variables for each individual and the data are cross-classified in a multidimensional contingency table. Various multivariate discrete distributions are then of interest (see Bishop *et al.*, 1975, Chapter 13), such as the multivariate form of the multinomial distribution (Johnson and Kotz, 1969, Section 11.9).

2.5.2 *Multivariate continuous distributions*

A whole book is devoted to this topic by Johnson and Kotz (1972). The most important distribution is the multivariate normal already considered in Section 2.3. Several distributions are related to the multivariate normal. These include the Wishart distribution (see Chapter 6) and the multivariate form of the t-distribution (see Johnson and Kotz, 1972, Chapter 37). Other distributions considered by Johnson and Kotz include multivariate forms of the beta, gamma, extreme-value and exponential distributions.

As an example, let us consider the multivariate beta distribution, which is often called the Dirichlet distribution. A random variable X is said to have a (univariate) *beta* distribution with parameters α and β if its p.d.f. is given by

$$f(x) = x^{\alpha-1}(1-x)^{\beta-1}/B(\alpha, \beta) \qquad 0 < x < 1$$

where $B(\alpha, \beta) = \int_0^1 x^{\alpha-1}(1-x)^{\beta-1}\mathrm{d}x$ denotes the beta function, and $\alpha > 0$, $\beta > 0$.

The random variables $X_1, \ldots, X_p$ are said to follow the *Dirichlet* distribution with parameters $\alpha_1, \ldots, \alpha_p$ if their joint p.d.f. is given by

$$f(x_1, \ldots, x_p) = \frac{\Gamma(C) \prod_{i=1}^{p} x_i^{\alpha_i - 1}}{\prod_{i=1}^{p} \Gamma(\alpha_i)}$$

where $x_i \geq 0$ for all i but the x's are subject to the constraint that $\sum_{i=1}^{p} x_i = 1$. In the above expression, $C = \sum_{i=1}^{p} \alpha_i$, $\alpha_i > 0$ for all i and $\Gamma(\alpha) = \int_0^\infty t^{\alpha-1}\mathrm{e}^{-t}dt$ denotes the gamma function. Note that the gamma and beta functions are related by $B(\alpha, \beta) = \Gamma(\alpha)\Gamma(\beta)/\Gamma(\alpha+\beta)$. The Dirichlet distribution has been used to describe the distribution of purchase probabilities for a population of individuals buying one and only one brand of a particular product (such as soap powder), as described by Chatfield and Goodhardt (1974). This sort of problem will not be pursued in this book.

Exercises

2.1 Three random variables, X, Y, Z, have joint p.d.f.

$$f(x, y, z) = \begin{cases} kxyz^2 & 0 < x, y < 1; 0 < z < 3 \\ 0 & \text{otherwise} \end{cases}$$

Show that the constant k is 4/9, and that the random variables are independent. Show that the marginal joint p.d.f. of X and Z is given by

$$f(x, z) = \begin{cases} 2xz^2/9 & 0 < x < 1, 0 < z < 3 \\ 0 & \text{otherwise} \end{cases}$$

Show that $E(X) = 2/3$. Find the conditional distribution of X given $Y = 1/2$, $Z = 1$. Check that the covariance of X and Z is zero.

2.2 Two random variables, X and Y, have joint p.d.f.

$$f(x, y) = \begin{cases} kx & 0 < y < x < 1 \\ 0 & \text{otherwise} \end{cases}$$

Evaluate the constant k. Obtain the marginal distributions of X and Y and show that the random variables are not independent.

2.3 Let ρ denote the correlation coefficient for two random variables, X and Y. Show that $|\rho| \leq 1$.
If X and Y are independent, show that $\rho = 0$. However, by considering the joint p.d.f.

$$f(x, y) = \exp[-\tfrac{1}{2}(x^2 + y^2)]/[\sqrt{2}\pi^{3/2}(x^2 + y^2)^{1/2}]$$
$$-\infty < x, y < +\infty,$$

show that $\rho = 0$ does not imply independence.
[*Hint*: You do *not* need to evaluate any nasty double integrals. Rather you should use the symmetry of $f(x, y)$.]

2.4 Suppose that X_1, X_2, X_3, are independent random variables with unit variance. Let $Y_1 = X_1 + X_2 + X_3$, $Y_2 = X_1 - X_2$, and $Y_3 = X_1 - X_3$. Find the covariance and correlation matrices of $\mathbf{Y}$, where $\mathbf{Y}^{\mathrm{T}} = [Y_1, Y_2, Y_3]$.

2.5 If $\mathbf{X}$ is a p-dimensional random variable with covariance matrix Σ and A is a $(p \times m)$ matrix of constants, show that the covariance matrix of $A^{\mathrm{T}}\mathbf{X}$ is given by $A^{\mathrm{T}}\Sigma A$. Check that this result can be applied to Exercise 2.4 above.

2.6 Let $\mathbf{X}^{\mathrm{T}} = [X_1, \ldots, X_p]$ be a p-variate random variable with covariance matrix Σ and correlation matrix P. Show that P is positive semidefinite. What further condition is necessary for Σ and P to be positive definite?
[*Hint*: Use Equations (2.11) and (2.8)].

2.7 Show that the correlation coefficient of two variables X and Y is invariant under linear transformations of either or both variables.

2.8 Suppose the joint distribution of two random variables, X_1 and X_2, is the bivariate normal distribution with parameters $\mu_1, \mu_2, \sigma_1^2, \sigma_2^2, \rho$, as described in Section 2.4. Find the inverse of the covariance matrix and the determinant of the covariance matrix.

What happens to the distribution when: (*a*) $\rho = 0$; (*b*) $\rho = +1$; (*c*) $\rho = -1$?

CHAPTER THREE

Preliminary data analysis

Having collected a set of multivariate data, the first step is to process the data and carry out a preliminary data analysis in order to get a general 'feel' of the data. These preliminary procedures are relatively straightforward, but it must be emphasized that they form an essential part of any multivariate analysis. With computer programs readily available, a deplorable tendency has grown up in recent years for researchers in some quarters to rush into using complicated multivariate techniques, like factor analysis, without first having a careful 'look' at the data.

3.1 Processing the data

Most data are recorded manually on data sheets. Unless the numbers of individuals and variables are small (e.g., $n < 100$ and $p = 2$), the data will probably be analysed on a computer. This means that the data must go through three stages:

(*a*) *Coding.* The data have to be transferred to coding sheets so that they can be punched up in an appropriate way.

(*b*) *Punching.* The data are recorded on cards, magnetic tape or disc.

(*c*) *Editing.* The data are checked for errors.

Any statistician carrying out a multivariate analysis must be prepared to tackle this important part of the operation. Yet surprisingly few textbooks even mention the subject. Here we will draw attention to some of the main points which need to be considered, and refer the reader to Naus (1975) and O'Muircheartaigh and Payne (1977, Chapter 2) for further details.

The first step in 'screening' the data is to decide if all the variables are worth including, or if some may be disregarded. Then, when coding the data, the statistician must consider the following points:

(*a*) Choice of input medium. For several years, most data sets were recorded on punched cards, but there is now increasing use of what are called secondary media, such as magnetic tape and disc. Recording the data on magnetic tape or disc requires the intervention of a computer process, but the data are then more efficient to handle on modern computer installations.

(*b*) Coding of non-numeric data. Many data sets include some variables which are not recorded in numerical form. Examples include opinion responses, which might range from 'agree strongly' to 'disagree strongly', and qualitative variables like 'colour of hair'. The analyst needs to code such variables with extra care. For *quantitative* variables, whose possible values can be arranged in order of magnitude, it is often convenient to code the data in numerical form even though some information may be lost in the process. Opinion responses are often coded on a five-point scale, with equally spaced values, so that 'agree strongly' and 'disagree strongly' might be coded as 1 and 5 respectively, with intermediate values as appropriate. It is more difficult to deal with *qualitative* variables, where the possible values cannot be arranged in any logical numerical order. For example if the individuals are recorded as 'male' or 'female', then we have a binary qualitative variable which could be coded as 1 for 'male' and 0 for 'female'. If there are more than two possible values, as for example if one records a man's hair colour as 'black', 'brown' or 'grey', then any numerical coding is arbitrary. But as many computer programs have difficulty in handling non-numeric data, some such coding may have to be done.

(*c*) Missing values. If any observations are missing for any reason, they must be carefully coded to distinguish them from ordinary observations. In particular, when data are collected by means of sample surveys, the coding must distinguish between 'refused to reply', 'don't know' and 'not applicable'. Missing values can occasionally be 'filled in' from other sources.

(*d*) Choice of format. The most appropriate format must be selected for each variable. It is usually safer to use a fixed-column format rather than a free-field format.

(*e*) Choice of order. The most sensible order for the different variables must also be selected.

If punched cards are to be used, the analyst must also consider if all the observations for one individual will go conveniently on to one card. Three pieces of advice are relevant here. Firstly, one should avoid trying to get more than one piece of information into one column by means of what are called multipunched codes. Secondly, one should leave at least one blank column between each observation. Thirdly, one should put an identifying label on each card in case the cards get dropped! Also on the subject of punched cards, we note that, if several individuals have the same characteristics for several of the variables, it may be sensible to put this common information on marker cards rather than try to duplicate it on each card. The data are then said to have a hierarchical structure.

After the data have been coded, they can be *punched* on to cards, tape or disc. The punch operator should be encouraged to call attention to any obvious mistakes on the coding sheets. Data on cards should be *verified* by repunching, although experience suggests that errors will often be repeated. At this stage, punched cards can be read into the computer and copied on to magnetic tape or disc.

3.1.1 *Data editing*

Having recorded the data, it is then essential to look for suspect values and errors of various kinds. There are many different types of suspect value and it is helpful to distinguish between them.

(*a*) *Outliers.* These are defined to be observations which appear to be inconsistent with the rest of the data. They may be caused by gross recording or punching errors. But it is important to realize that an apparent outlier may occasionally be genuine and indicate a non-normal distribution.

(*b*) *Inversions.* A common type of error occurs when two successive digits are interchanged at the recording, coding or punching stage. The error may be trivial if, for example, 123.45 appears as 123.54, but it may produce an outlier if 123.45 appears as 213.45.

(*c*) *Repetitions.* At the coding or punching stage, it is quite easy to repeat a whole number in two successive rows or columns of a table, thereby omitting one number completely.

(*d*) *Values in the wrong column.* It is also easy to get numbers into the wrong columns.

(*e*) *Other errors and suspect values.* There are many other types of error including possible misrecordings of a trivial kind.

The general term used to denote procedures for detecting and correcting errors is *data editing*. This includes checks for *completeness, consistency* and *credibility*. Some editing can be done at the coding stage. In addition, many routine checks can be made by the computer itself, particularly those for gross outliers. An important class of such checks are *range tests*. For each variable an allowable range of possible values is specified, and the computer checks that all observed values lie within the given range. Bivariate and multivariate checks are also possible. For example, one may specify an allowable range for some function of two or more variables. A set of checks called 'if–then' checks are also possible. For example if both age and date of birth are recorded for each individual, then one can check that the answers are consistent. If the date of birth is given, then one can deduce the corresponding age. In fact, in this example the age observation is *redundant*. It is sometimes a good idea to include one or two redundant variables as a check on accuracy. Various other general procedures for detecting outliers are described by Barnett and Lewis (1978, Chapter 6). Some of these methods depend on plotting the data in various ways and looking subjectively for outlying observations, as described more fully in Section 3.3.

In addition to the above procedures, another simple but very useful check is to get a printout of the data and look at it by eye. Although it may be impractical to check every digit visually, the eye is very efficient at picking out many types of obvious error, particularly repetitions and gross outliers. Table 3.1 shows a 'typical' printout of a set of data with 5 observations on 4 variables at each of 3 levels. The data contains several obvious suspect values which have been encircled.

Table 3.1

	Variable			
	1	2	3	4
Level 1	0.0	103.1	93.2	23
	0.0	110.2	87.2	27
	0.0	110.2	88.9	49
	0.0	105.8	92.1	24
	0.0	107.8	84.4	26
Level 2	2.1	87.4	117.1	13
	2.4	83.3	125.8	12
	2.5	87.2	132.1	10
	2.2	85.0	85.0	12
	2.2	89.0	126.5	12
Level 3	4.7	48.6	140.2	6
	14.8	44.2	145.5	6
	4.6	49.3	138.7	5
	5.0	49.7	193.2	6
	4.7	40.1	142.2	6

The repetition of the value 110.2 at level 1 is suspicious but could conceivably be correct. The repetition of values of variables 1 and 4 are not suspicious in view of their smaller variation. The repetition across columns of the value 85.0 at level 2 is far more suspicious as it gives rise to an apparent outlier in variable 3. At level 3, the value of 14.8 appears to have a spurious digit 'one' added to 4.8, while the outlier 193.2 may have inverted digits from 139.2. The outlier, 49, at level 1 has no obvious explanation.

When a possible suspect value or error has been detected, the statistician must decide what to do about it. One may be able to go back to the original data source and check the observation. Inversions, repetitions and values in the wrong column can often be corrected in this way. Outliers are more difficult to handle, particularly when they are impossible to check or have been misrecorded in the first place. It may be sensible to treat them as missing values and try to insert a value 'guessed' in an appropriate way (e.g., by interpolation or by prediction from other variables). Alternatively, the value may have to be left as unrecorded and then either all observations for the given individual will have to be discarded or one will have to accept unequal numbers of observations for the different variables.

The key point to realize here is that errors are unavoidable and that steps must be taken to deal with them as far as possible. As an example, Naus (1975) quotes the US Census of Population and Housing as involving 50 million omissions and inconsistencies in the 2 billion data fields processed. As another example, the director of a national data bank has stated that up to 15 per cent of the individual records will typically contain at least one error. Unfortunately, a few data errors can sometimes ruin the ensuing analysis. As a cautionary tale, the authors recall analysing some data on road accidents which were supplied ready-punched. Some peculiar relationships with age were found until it was realized that some lady drivers had omitted to give their age and had all been misleadingly coded as '99'. A simple range test would have spotted this. The moral is that one should never trust a large set of data to be correct.

3.2 Calculating summary statistics

With a univariate set of observations, the analysis usually begins with the calculation of two summary statistics, namely the mean and standard deviation. In the multivariate case, the analysis usually begins with the calculation of the mean and standard deviation for each variable, and in addition the correlation coefficient for each pair of variables is usually calculated. These summary statistics are vital in having a preliminary look at the data.

The *sample mean* of the jth variable is given by

$$\bar{x}_j = \sum_{r=1}^{n} x_{rj} \Big/ n \tag{3.1}$$

and the sample mean vector, $\bar{\mathbf{x}}$, is given by $\bar{\mathbf{x}}^T = [\bar{x}_1, \bar{x}_2, \ldots, \bar{x}_p]$. If the observations are a random sample from a population with mean $\boldsymbol{\mu}$, then the sample mean vector $\bar{\mathbf{x}}$ is the usual point estimate of $\boldsymbol{\mu}$, and this estimate can easily be shown to be unbiased.

The standard deviation of the jth variable is given by

$$s_j = \sqrt{\left[\sum_{r=1}^{n} (x_{rj} - \bar{x}_j)^2 \Big/ (n-1)\right]}$$

The correlation coefficient of variables i and j is given by

$$r_{ij} = \sum_{r=1}^{n} (x_{ri} - \bar{x}_i)(x_{rj} - \bar{x}_j) \Big/ (n-1)s_i s_j.$$

These coefficients can be conveniently assembled in the sample correlation matrix, R, which is given by

$$R = \begin{bmatrix} 1 & r_{12} & \cdots & r_{1p} \\ r_{21} & 1 & \cdots & r_{2p} \\ \vdots & & & \\ r_{p1} & r_{p2} & \cdots & 1 \end{bmatrix}$$

Note that the diagonal terms are all unity. This matrix provides an estimate of the corresponding population correlation matrix, P, which was defined in Section 2.2. We note in passing that this estimate is generally *not* unbiased, but the bias is generally small and does not stop us using the estimate. The virtues of unbiasedness are sometimes overstressed.

The interpretation of means and standard deviations is straightforward. It is worth looking to see if, for example, some variables have much higher scatter than others. It is also worth looking at the form of the distribution of each variable, and considering whether any of the variables need to be transformed. For example, the logarithmic transformation is often used to reduce positive skewness and produce a distribution which is closer to normal. One may also consider the removal of outliers at this stage (see also Sections 3.1 and 3.3).

The interpretation of the sample correlation matrix is discussed in detail in Section 3.2.1. But first we will discuss the computation of R via the sample covariance matrix.

The sample covariance matrix. The sample variances, standard deviations and correlations are usually calculated from the elements of the sample covariance matrix S, whose (i, j)th element is usually defined as

$$\begin{aligned} s_{ij} &= \text{sample covariance of variables } i \text{ and } j \\ &= \sum_{r=1}^{n} (x_{ri} - \bar{x}_i)(x_{rj} - \bar{x}_j) \Big/ (n-1) \end{aligned} \tag{3.2}$$

The denominator $(n-1)$ is usually chosen in preference to n so that s_{ij} turns out to be an unbiased estimate of the corresponding population covariance denoted by σ_{ij}. The proof is left as an exercise for the reader.

Putting $j = i$ in Equation (3.2), we get the familiar formula for the sample variance of the ith variable. This is usually denoted by s_{ii} in the multivariate case, rather than the more familiar s_i^2 which would be used in the univariate case. This makes it easier to write down the sample covariances in the matrix form

$$S = \begin{bmatrix} s_{11} & s_{12} & \cdots & s_{1p} \\ s_{21} & s_{22} & \cdots & s_{2p} \\ \vdots & & & \\ s_{p1} & s_{p2} & \cdots & s_{pp} \end{bmatrix}$$

As s_{ij} is unbiased for σ_{ij} for every i and j, we say that S is an unbiased estimator for the population covariance matrix Σ which was defined in Section 2.2. S is clearly symmetric and its diagonal terms give the variances of the variables. The standard deviations are then given by $s_i = \sqrt{s_{ii}}$. The off-diagonal terms of S are occasionally of interest in their own right. But the quantities depend on the scales used to measure the variables and are generally difficult to interpret as a measure of association between a given pair of variables. Thus they are usually used only as stepping stones in the calculation of sample correlation coefficients by

$$r_{ij} = s_{ij}/s_i s_j \tag{3.3}$$

These coefficients can be thought of as normalized sample covariances which are arranged to lie between -1 and $+1$.

The elements of S may be calculated directly using Equation (3.2). The equivalent matrix expression for S in terms of the data matrix X is

$$S = (X - \mathbf{1}\bar{\mathbf{x}}^{\mathrm{T}})^{\mathrm{T}}(X - \mathbf{1}\bar{\mathbf{x}}^{\mathrm{T}})/(n-1) \tag{3.4}$$

where $\mathbf{1}$ denotes an $(n \times 1)$ vector of ones. The equivalent mean-corrected version of Equation (3.4), namely

$$S = (X^{\mathrm{T}}X - n\bar{\mathbf{x}}\bar{\mathbf{x}}^{\mathrm{T}})/(n-1) \tag{3.5}$$

is not recommended as it is likely to give rise to rounding errors unless the data are suitably coded. But perhaps the most efficient way of calculating S is by means of the updating procedure described by Welford (1962).

If we define the diagonal matrix, $\hat{D}$, by

$$\hat{D} = \begin{bmatrix} s_1 & 0 & \dots & 0 \\ 0 & s_2 & \dots & 0 \\ \vdots & & & \\ 0 & 0 & \dots & s_p \end{bmatrix}$$

then, by analogy with Equation (2.11), we may calculate R from S using the formula

$$R = \hat{D}^{-1}S\hat{D}^{-1} \tag{3.6}$$

3.2.1 *Interpreting the sample correlation matrix*

We assume the reader is reasonably familiar with the interpretation of a single correlation coefficient (see for example Chatfield, 1978, Section 8.9). In particular, we recall that a correlation coefficient close to $+1$ (or -1) indicates a strong linear relationship with a positive (or negative) slope between the two variables concerned, while a coefficient close to zero indicates no linear relationship. We emphasize the word 'linear' here and remind the reader that it is generally unwise to calculate the correlation coefficient for a pair of variables whose relationship is obviously non-linear. In particular, zero correlation does not of course imply independence.

It is sometimes useful to test which coefficients are significantly different from zero. The test assumes that each pair of variables follows a bivariate normal distribution. A correlation r is significantly different from zero if the absolute value of $r\sqrt{(n-2)}/\sqrt{(1-r^2)}$ exceeds the appropriate percentage point of the t-distribution with $(n-2)$ degrees of freedom. But when n is large, nearly all the coefficients will be 'significant', and estimation becomes more important. In particular, it is useful to assess which coefficients are large enough to be of practical importance. In this context it is helpful to recall the connection between correlation and residual variance. If ρ denotes the correlation between two variables y and x, then ρ^2 denotes the proportion of the total corrected sum of squares of y 'explained' by the linear regression of y on x, and the residual variance, $\sigma^2_{y|x}$, is related to the unconditional variance of y, σ^2_y, by $\sigma^2_{y|x} = \sigma^2_y(1-\rho^2)$. Thus, if for example $\rho = 0.6$, then only 36 per cent of the total sum of squares is explained by the regression and the residual standard deviation of y is still 80 per cent of the original standard deviation. So, as a rough guide, we suggest that values bigger than about 0.7 may be considered 'large'.

Note that correlations should generally not be calculated at all if the number of observations is too small to give reliable estimates (e.g., $n <$ about 12).

After these preliminary remarks, we turn now to the problems which face the analyst on being confronted with a $(p \times p)$ sample correlation matrix containing $p(p-1)/2$ distinct coefficients. The first difficulty to overcome is that computer output is usually in a totally unsuitable format for the purposes of interpretation. Too many significant figures are usually given so that all the numbers need to be rounded. In addition, if there are more than about seven variables, the output is often of the 'folding-back' type where one 'row' of the correlation matrix is actually spread over two or more rows in the computer output. This may be fun for the programmer but is no good for the user.

As an example, consider the data analysed by Ahamad (1967), which show the number of criminal offences of eighteen different types for Great Britain for the years 1950–63. The data are given in Table 3.2. Since $n < p$, where $n =$ 14 (years) and $p = 18$ (types of crime), it is convenient to give the data matrix in its transposed form.

The correlation matrix which results from one typical computer 'package' is shown in Table 3.3, together with a 'label' for each variable.

The coefficients are given with no less than seven decimal places, and the bottom row of the matrix is spread over three rows of computer output. Values above the diagonal do not have to be given as the matrix is symmetric. But the table is clearly impossible to interpret in the given form.

Various rules for improving the presentation of tables are given by Ehrenberg (1975, 1977). Although apparently self-evident, they are often ignored in practice, and so we give the following guidance to help in making sense of a correlation matrix.

(*a*) Round the coefficients to two decimal places. It is unlikely that one will ever need greater accuracy. Indeed, it is sometimes clearer to round to one decimal place provided that the second decimal place can be stored accessibly.

(*b*) Suppress the diagonal terms which are all unity, as they distract the eye without giving useful information.

(*c*) Suppress the 0's which appear before the decimal point in all off-diagonal coefficients.

(*d*) Although the matrix is symmetric, it is helpful to give the values above the diagonal as well as below.

(*e*) Put the rows and columns reasonably close together, as the numbers are easier to compare if the eye does not have to move around too much. But leave a slightly larger gap between the rows (and the columns) after about every five rows (and after about every five columns). This makes it easier to locate any particular variable.

Table 3.2 Recorded numbers of offences of 18 types of crime, 1950–63

Variate	1950	1951	1952	1953	1954	1955	1956	1957	1958	1959	1960	1961	1962	1963
X_1	529	455	555	456	487	448	477	491	453	434	492	459	504	510
X_2	5 258	5 619	5 980	6 187	6 586	7 076	8 433	9 774	10 945	12 707	14 391	16 197	16 430	18 655
X_3	4 416	4 876	5 443	5 680	6 357	6 644	6 196	6 327	5 471	5 732	5 240	5 605	4 866	5 435
X_4	8 178	9 223	9 026	10 107	9 279	9 953	10 505	11 900	11 823	13 864	14 304	14 376	14 788	14 722
X_5	92 839	95 946	97 941	88 607	75 888	74 907	85 768	105 042	131 132	133 962	151 378	164 806	192 302	219 138
X_6	1 021	800	1 002	980	812	823	965	1 194	1 692	1 900	2 014	2 349	2 517	2 483
X_7	301 078	355 407	341 512	308 578	285 199	295 035	323 561	360 985	409 388	445 888	489 258	531 430	588 566	635 627
X_8	25 333	27 216	27 051	27 763	26 267	22 966	23 029	26 235	29 415	34 061	36 049	39 651	44 138	45 923
X_9	7 586	9 716	9 188	7 786	6 468	7 016	7 215	8 619	10 002	10 254	11 696	13 777	15 783	17 777
X_{10}	4 518	4 993	5 003	5 309	5 251	2 184	2 559	2 965	3 607	4 083	4 802	5 606	6 256	6 935
X_{11}	3 790	3 378	4 173	4 649	4 903	4 086	4 040	4 689	5 376	5 598	6 590	6 924	7 816	8 634
X_{12}	118	74	120	108	104	92	119	121	164	160	241	205	250	257
X_{13}	20 844	19 963	19 056	17 772	17 379	17 329	16 677	17 539	17 344	18 047	18 801	18 525	16 449	15 918
X_{14}	9 477	10 359	9 108	9 278	9 176	9 460	10 997	12 817	14 289	14 118	15 866	16 399	16 852	17 003
X_{15}	24 616	21 122	23 339	19 919	20 585	19 197	19 064	19 432	24 543	26 853	31 266	29 922	34 915	40 434
X_{16}	49 007	55 229	55 635	55 688	57 011	57 118	63 289	71 014	69 864	69 751	74 336	81 753	89 709	89 149
X_{17}	2 786	2 739	2 598	2 639	2 587	2 607	2 311	2 310	2 371	2 544	2 719	2 820	2 614	2 777
X_{18}	3 126	4 595	4 145	4 551	4 343	4 836	5 932	7 148	9 772	11 211	12 519	13 050	14 141	22 896

Applying these rules to Table 3.3, we get Table 3.4. The table still looks rather daunting and needs to be studied carefully for several minutes. Yet it is an unfortunate fact that people are often too lazy to do this. For some reason they think it is easier to get the computer to carry out more sophisticated analyses, like principal component analysis (see Chapter 4), even though it may well take longer even to get access to a computer terminal than to carry out a careful visual inspection of the matrix.

In studying a large table like Table 3.4, it is a good idea to take a couple of photocopies so that one can 'mess it about'. In particular, it may be a good idea to draw rings round 'large' correlations (e.g., those whose absolute value exceeds 0.7) and cross out 'small' correlations (e.g., those whose absolute value is less than about 0.25).

Close inspection of Table 3.4 reveals some clear patterns. For example, the correlations with variable 3 are all negative. This would not be at all obvious if the values above the diagonal were omitted. We also see that many of the correlations are close to ± 1. For example, in the second row we see that variable 2 is highly correlated ($r \geq 0.88$) with variables 4,5,6,7,8,9,11,12,14,15,16,18. Looking back at Table 3.2, we see that crimes in all these categories have increased steadily over the years and simply reflect the 'crime explosion'. The other variables need to be treated separately. Variable 1 (homicide) has low correlation with all other variables as its recorded incidence has changed relatively little over the years. The negative correlations with variable 3 (homosexual offences) arise because the incidence actually decreased in the last few years when the other variables were increasing at their fastest rates. It is straightforward to make similar interpretations for the changes in variables 10, 13 and 17. In fact, the changes in some of the latter variables are clearly non-linear and it is debatable whether correlations should be calculated for every pair of variables.

Of course these data are rather unusual in that they are time-series data and there are relatively few observations ($n = 14$), so that the patterns are nearly as obvious in the original data matrix as they are in the correlation matrix. (The example was chosen for that reason and will be further used to illustrate principal component analysis in Chapter 4.) When n is larger and the data are not time-series, the correlation matrix will usually be invaluable.

The pattern of behaviour in a large correlation matrix may become clearer if the variables are *re-ordered* to bring out the natural groupings. Thus, in Table 3.4 the natural order would probably be the highly correlated group of variables, followed by variables 10 and 17, then variable 1, and finally variables 3 and 13. However, in this case re-ordering gives only a marginal improvement in clarity. A much more convincing example is given by Ehrenberg (1977).

Table 3.5 shows the correlations between ten British TV programmes like *Professional Boxing* (PrB), *This Week* (ThW), and so on, for 7000 UK adults who were asked if they 'really liked to watch each programme'.

The patterns, if any, in the table are unclear. But by rounding the

Table 3.3 The correlation matrix for the data in Table 3.2

Variate		Homicide 1	Wounding 2	Homosex 3	Heterosex 4	Brk-entr 5	Robbery 6	Larceny 7
Homicide	1	1.0000000						
Wounding	2	−0.0406881	1.0000000					
Homosex	3	−0.3709117	−0.1329193	1.0000000				
Heterosex	4	−0.1744020	0.9693951	−0.0704008	1.0000000			
Brk-entre	5	0.1283502	0.9471402	−0.3790885	0.8813339	1.0000000		
Robbery	6	0.0130596	0.9698411	−0.3150878	0.9403094	0.9633480	1.0000000	
Larceny	7	0.0742331	0.9610246	−0.3386777	0.9094577	0.9922534	0.9659860	1.0000000
Fraud	8	0.0989100	0.9229339	−0.3765394	0.8696565	0.9736497	0.9502706	0.9751580
Rec-stol	9	0.1571127	0.9002824	−0.4064417	0.8223439	0.9813841	0.9151790	0.9815479
Inj-prop	10	0.3467703	0.4689174	−0.5520135	0.3714722	0.6483568	0.5504609	0.6233604
Forgery	11	0.0891201	0.9535277	−0.1731940	0.8962558	0.9476328	0.9415627	0.9415404
Blckmail	12	0.1827698	0.9413795	−0.3223804	0.8947414	0.9427847	0.9561235	0.9358546
Assault	13	0.1734503	−0.5015650	−0.5339872	−0.4878591	−0.3847117	−0.3670331	−0.3893046
Mal-damg	14	−0.0815110	0.9724516	−0.2339077	0.9619668	0.9297636	0.9636167	0.9436036
Revenue	15	0.2640710	0.8758820	−0.4705026	0.7817415	0.9612115	0.9095209	0.9427009
Alcohol	16	−0.0143671	0.9728866	−0.0864408	0.9461783	0.9258120	0.9326415	0.9420567
Indecent	17	0.1733367	0.2014737	−0.5454496	0.1092685	0.3390750	0.2824276	0.3350558
Motor	18	0.0150658	0.9567783	−0.1667589	0.8931665	0.9586458	0.9159223	0.9543911

Variate		Variate						
		Fraud 8	Rec-stol 9	Inj-prop 10	Forgery 11	Blckmail 12	Assault 13	Mal-damg 14
Fraud	8	1.0000000						
Rec-stol	9	0.9673345	1.0000000					
Inj-prop	10	0.7508285	0.7009911	1.0000000				
Forgery	11	0.9571994	0.9093160	0.6446658	1.0000000			
Blckmail	12	0.9223379	0.8868445	0.5650396	0.9504277	1.0000000		
Assault	13	−0.3537935	−0.3464847	−0.0267828	−0.5353122	−0.3886208	1.0000000	
Mal-damg	14	0.8854982	0.8750712	0.4120566	0.8934680	0.9216996	−0.3973719	1.0000000
Revenue	15	0.9536476	0.9445976	0.7228256	0.9196069	0.9319863	−0.2717484	0.8344792
Alcohol	16	0.8949883	0.8930854	0.4447675	0.9313692	0.9005131	−0.5920886	0.9550186
Indecent	17	0.4482233	0.4140759	0.6772028	0.3040931	0.2858098	0.4276267	0.1439093
Motor	18	0.9194168	0.9256895	0.5420980	0.9436258	0.9049507	−0.522663	0.9094464

Variate		Variate			
		Revenue 15	Alcohol 16	Indecent 17	Motor 18
Revenue	15	1.0000000			
Alcohol	16	0.8171265	1.0000000		
Indecent	17	0.4983916	0.0731388	1.0000000	
Motor	18	0.9146270	0.9202035	0.2559079	1.0000000

Table 3.4 The revised correlation matrix

Variable	1	2	3	4	5	6	7	8	9	10	11	12	13	14	15	16	17	18
1		−.04	−.37	−.17	.13	.01	.07	.10	.16	.35	.09	.18	.17	−.08	.26	−.01	.17	.02
2	−.04		−.13	.97	.95	.97	.96	.92	.90	.47	.95	.94	−.50	.97	.88	.97	.20	.96
3	−.37	−.13		−.07	−.38	−.31	−.34	−.38	−.41	−.55	−.17	−.32	−.53	−.23	−.47	−.09	−.55	−.17
4	−.17	.97	−.07		.88	.94	.91	.87	.82	.37	.90	.89	−.49	.96	.78	.95	.11	.89
5	.13	.95	−.38	.88		.96	.99	.97	.98	.65	.95	.94	−.38	.93	.96	.93	.34	.96
6	.01	.97	−.31	.94	.96		.97	.95	.92	.55	.94	.96	−.37	.96	.91	.93	.28	.92
7	.07	.96	−.34	.91	.99	.97		.98	.98	.62	.94	.94	−.39	.94	.94	.94	.33	.95
8	.10	.92	−.38	.87	.97	.95	.98		.97	.75	.96	.92	−.35	.89	.95	.89	.45	.92
9	.16	.90	−.41	.82	.98	.92	.98	.97		.70	.91	.89	−.35	.88	.94	.89	.41	.93
10	.35	.47	−.55	.37	.65	.55	.62	.75	.70		.64	.57	−.03	.41	.72	.44	.68	.54
11	.09	.95	−.17	.90	.95	.94	.94	.96	.91	.64		.95	−.54	.89	.92	.93	.30	94
12	.18	.94	−.32	.89	.94	.96	.94	.92	.89	.57	.95		−.39	.92	.93	.90	.29	.90
13	.17	−.50	−.53	−.49	−.38	−.37	−.39	−.35	−.35	−.03	−.54	−.39		−.40	−.27	−.59	.43	−.52
14	−.08	.97	−.23	.96	.93	.96	.94	.89	.88	.41	.89	.92	−.40		.83	.96	.14	.91
15	.26	.88	−.47	.78	.96	.91	.94	.95	.94	.72	.92	.93	−.27	.83		.82	.50	.91
16	−.01	.97	−.09	.95	.93	.93	.94	.89	.89	.44	.93	.90	−.59	.96	.82		.07	.92
17	.17	.20	−.55	.11	.34	.28	.33	.45	.41	.68	.30	.29	.43	.14	.50	.07		.26
18	.02	.96	−.17	.89	.96	.92	.95	.92	.93	.54	.94	.90	−.52	.91	.91	.92	.26	

Table 3.5 Correlations for ten TV programmes
(The programmes are ordered alphabetically within channel)

		PrB	ThW	Tod	WoS	GrS	LnU	MoD	Pan	RgS	24H
ITV	PrB	1.0000	0.1064	0.0653	0.5054	0.4741	0.0915	0.4732	0.1681	0.3091	0.1242
	ThW	0.1064	1.0000	0.2701	0.1424	0.1321	0.1885	0.0815	0.3520	0.0637	0.3946
	Tod	0.0653	0.2701	1.0000	0.0926	0.0704	0.1546	0.0392	0.2004	0.0512	0.2437
	WoS	0.5054	0.1424	0.0926	1.0000	0.6217	0.0785	0.5806	0.1867	0.2963	0.1403
BBC	GrS	0.4741	0.1321	0.0704	0.6217	1.0000	0.0849	0.5932	0.1813	0.3412	0.1420
	LnU	0.0915	0.1885	0.1546	0.0785	0.0849	1.0000	0.0487	0.1973	0.0969	0.2661
	MoD	0.4732	0.0815	0.0392	0.5806	0.5932	0.0487	1.0000	0.1314	0.3267	0.1221
	Pan	0.1681	0.3520	0.2004	0.1867	0.1813	0.1973	0.1314	1.0000	0.1469	0.5237
	RgS	0.3091	0.0637	0.0512	0.2963	0.3412	0.0969	0.3267	0.1469	1.0000	0.1212
	24H	0.1242	0.3946	0.2437	0.1403	0.1420	0.2661	0.1211	0.5237	0.1212	1.0000

correlations to one decimal place, clearly labelling the variables, arranging better spacing in the table, and by appropriate re-ordering of the variables, the table becomes much easier to understand. In particular, the cluster of five correlated sports programmes is easy to see, as are the generally low correlations between sports programmes and current affairs programmes.

Choosing the right order for a set of variables is not always easy. One relies mainly on a visual inspection of the correlation matrix together with the use of prior information, such as that sports programmes may tend to go together. Of course prior knowledge could be used to input the variables in a sensible order in the first place by grouping variables which are expected to be highly correlated.

Having decided upon a suitable order for the variables, it is physically rather tedious to actually do the re-ordering of the rows and columns of a correlation matrix, as has been done in Table 3.6. While a computer program could be written to do this, it is probably easiest to do by hand. Trying to transcribe the numbers in writing often leads to errors, so the best method is physically to re-

Table 3.6 The correlations in Table 3.5 rounded and re-ordered

Programmes		WoS	MoD	GrS	PrB	RgS	24H	Pan	ThW	Tod	LnU
World of Sport	ITV		.6	.6	.5	.3	.1	.2	.1	.1	.1
Match of the Day	BBC	.6		.6	.5	.3	.1	.1	.1	0	0
Grandstand	BBC	.6	.6		.5	.3	.1	.2	.1	.1	.1
Prof. Boxing	ITV	.5	.5	.5		.3	.1	.2	.1	.1	.1
Rugby Special	BBC	.3	.3	.3	.3		.1	.1	.1	.1	.1
24 Hours	BBC	.1	.1	.1	.1	.1		.5	.4	.2	.3
Panorama	BBC	.2	.1	.2	.2	.1	.5		.4	.2	.2
This Week	ITV	.1	.1	.1	.1	.1	.4	.4		.3	.2
Today	ITV	.1	0	.1	.1	.1	.2	.2	.3		.2
Line-Up	BBC	.1	0	.1	.1	.1	.3	.2	.2	.2	

order the columns and then the rows in two separate stages, by cutting and pasting a photocopy of the table. First cut the table by columns and paste them onto a blank sheet of paper in the correct order. Then repeat this procedure on the columns.

To close this section, we repeat our advice that in order to 'make sense' of a correlation matrix, it is worth spending time getting the table into a clear format and then studying the elements of the matrix in detail.

3.2.2 *The rank of R*

We complete this section with a more technical discussion of the matrix properties of R. Firstly we note that S is positive semidefinite (see Exercise 3.2), and hence using Equation (3.6) we see that R is also positive semidefinite. Furthermore, as $\hat{D}$ is diagonal and clearly non-singular, we may use Equations (1.8) and (3.6) to show that the ranks of S and R must be identical. And, using Equations (1.7) and (3.4), we can show that this rank is also identical to the rank of the mean-corrected data matrix, namely $(X - \mathbf{1}\bar{\mathbf{x}}^T)$. As the latter matrix is of order $(n \times p)$, its rank must be less than or equal to min (n, p). Thus if there are less observations than variables, so that $n < p$, then rank $(X - \mathbf{1}\bar{\mathbf{x}}^T) < p$ and it follows that S and R will both be singular. In addition, if $n = p$, $(X - \mathbf{1}\bar{\mathbf{x}}^T)^T$ is a square matrix with a zero eigenvalue, as the sum of the values in each row is zero, so that its rank is strictly less than p. Then S and R will also be singular with rank less than p. To illustrate this last result, suppose $p = 3$ and we take only $n = 3$ observations. Then we can always fit a plane through three observations in three dimensions, so that the data are essentially two-dimensional. To avoid singularities arising from too small a sample, we would generally like n to be greater than p, preferably much greater, and in future we will often assume that this is the case.

If $n > p$, the rank of R or equivalently the rank of S or $(X - \mathbf{1}\bar{\mathbf{x}}^T)$ will generally be equal to p. But if there are linear constraints on the random variables, the constraints should also be present in the sample data leading to a singular mean-corrected data matrix. Then S and R will also be singular. It turns out that the number of independent linear constraints is equal to $p - \text{rank}(R)$ so that the rank of R is then a useful guide to the effective dimensionality of the data.

On the other hand if there are linear constraints on the *expected values* of the random variables, there will *not* be exact linear constraints on the sample data, though the determinant of R (or of S) may be 'small'. Indeed, even if there are exact linear constraints on the random variables (rather than their expected values), the determinant of R (or of S) will usually not vanish completely because of rounding errors, though the determinant may be very small. In this case the matrix is said to be *ill-conditioned* and it may be advisable to remove some of the variables, or to replace some or all of the variables by a smaller number of new linear combinations of the variables. If the matrix has k 'very small' eigenvalues, the dimensionality may be reduced from p to $(p - k)$ [see also Sections 4.3.3 and 4.3.4 and Section 8.3].

3.3 Plotting the data

It is always a good idea to plot data in whatever way seems appropriate. This should help the analyst get a 'feel' for his data, and may suggest relationships between the variables. It may also help the analyst to spot subjectively any outlying observations, to detect any natural clustering of the observations (see Section 11.2), and to check on distributional assumptions.

With just two variables ($p = 2$), each bivariate observation can be represented as a point on a two-dimensional graph, called a *scatter diagram*. The reader should already be familiar with diagrams of this type which are used in regression to see what sort of regression curve is appropriate to describe the relationship between two variables.

Three-dimensional data can be presented in several ways. One approach is to plot each observation as a circular 'blob' on a two-dimensional graph. The co-ordinates of the blob represent two of the variables, while its thickness represents the third variable. Alternatively, a three-dimensional model can be constructed using, for example, pins of different heights. Examples of both approaches are given by Marriott (1974, Figs. 1.1 and 1.3).

With more than three variables, pictorial representation of the data by the above direct type of approach becomes impractical, and alternative ways must be found. One simple approach is to plot the variables two at a time, so that with p variables one gets a total of $p(p-1)/2$ scatter diagrams. These diagrams may show up obvious bivariate relationships and obvious outliers, but are sometimes rather misleading as they consist of projections onto various planes, and it can be shown that the form of a multivariate distribution is not necessarily revealed by its bivariate marginal distributions. Another problem is that the number of scatter diagrams to be inspected becomes rather large if p is greater than about seven.

A completely different approach has been suggested by Andrews (1972), who suggests representing each p-variate observation by a function plotted over the range $(-\pi, +\pi)$. For the rth observation this function is defined by

$$f_r(t) = x_{r1}/\sqrt{2} + x_{r2}\sin t + x_{r3}\cos t + x_{r4}\sin 2t + x_{r5}\cos 2t + \ldots$$

Apart from the first term, this function is a mixture of sine and cosine waves, and will produce some sort of wave pattern depending on the observed values of the p variables. Observations which are 'close together' in p-dimensional space should give wave patterns which are somewhat similar. Indeed, Andrews (1972) shows that if the 'distance' between two functions is defined in the obvious way by

$$\int_{-\pi}^{\pi} [f_r(t) - f_s(t)]^2 \, dt$$

then this is proportional to the squared Euclidean distance between $\mathbf{x}_r$ and $\mathbf{x}_s$. The Andrews curves will not show up relationships between variables, but do appear to be useful for finding clusters and outliers. Some convincing examples are given by Everitt (1978, Chapter 4) and Jones (1979), and the method seems well worth considering.

One drawback to the method is that the results depend on the order in which the variables are labelled. If some of the variables are thought to be more important than others, then it is advisable to take the most important variable as x_1, and so on. Alternatively, particularly if the number of variables is rather high, the method can be applied not to the original data but to a set of transformed variables called principal components, which will be introduced in Chapter 4. The first few important components may be analysed, with the first (and most important) component labelled as x_1, and so on.

An alternative approach to the use of Andrews curves has been suggested by Chernoff (1973). His idea is to represent each p-variate observation by a cartoon face in two dimensions. Each face has a list of features which may be varied, such as the size of the mouth and the slope of the eyes. Each observed variable is made to correspond to one of these features. Some examples are given by Everitt (1978, Chapter 4) and Jones (1979). The technique has considerable amusement value, but in our view is not to be recommended. It is much harder to program than Andrews' method and it appears more difficult to compare a set of faces than a set of Andrews' curves. In particular, it is difficult to avoid subjectively giving more weight to some facial features than others.

Some other pictorial techniques for obtaining a direct representation of a set of multivariate data include the use of *glyphs* and *weathervane plots* (see Gnanadesikan, 1977, Chapter 3), but they will not be discussed here.

During the course of the book we shall describe a variety of more sophisticated techniques which indirectly provide simple graphical representations of the data. Many of these methods are essentially concerned with reducing the effective dimensionality of the data. Techniques for representing p-dimensional data in a much smaller number of dimensions, preferably 2 or 3, are often called *ordination* methods (see Everitt, 1978, Chapter 2). The methods include principal component analysis (Chapter 4) and multidimensional scaling (Chapter 10). In particular, we will see that plotting the values of the first two principal components gives a useful two-dimensional graph (as in Fig. 4.1) while classical scaling or ordinal scaling can be used to give a 'map' in two dimensions (as in Figs 10.1 and 10.3).

One final class of graphical procedures we will briefly mention are *probability plotting* methods, which, as in the univariate case, can be used to check distributional assumptions, to obtain rough estimates of distribution parameters, and to check for outliers. In the univariate case, the observations are ordered and then plotted against their expected values or against the appropriate values of the cumulative distribution function. In the multivariate case, we either plot one variable at a time or preferably convert the multivariate response to a single variable in an appropriate way. Thus in assessing multivariate normality, we can find a quantity called the *generalized distance* of each observation from the overall sample mean vector. For an observation $\mathbf{x}$, this is given by $(\mathbf{x}-\bar{\mathbf{x}})^T S^{-1}(\mathbf{x}-\bar{\mathbf{x}})$ and will have an approximate χ^2 distribution with p degrees of freedom. After ordering, the observed generalized distances are plotted against the order statistics of the appropriate χ^2 distribution. Departures from multivariate normality should be indicated by departures from linearity in the plot. The reader is referred to Everitt (1978, Chapter 4) and Gnanadesikan (1977, Chapter 6) for a detailed discussion of probability-plotting methods.

One use of graphs is to look for outlying observations. In some ways these are harder to find than in the univariate case (see Gnanadesikan, 1977, p. 271). To start with, it is difficult to define an outlier in the multivariate case. A simple univariate outlier may typically be thought of as 'one that sticks out at the end', but no such simple concept suffices in higher dimensions. For example, a vector response may be faulty because of a gross error in one of its components or because of systematic small errors in all of its components. Another problem is that a multivariate outlier may distort correlations as well as means and variances. Further details regarding the detection of multivariate outliers are given by Barnett and Lewis (1978, Chapter 6).

3.4 The analysis of discrete data

Many multivariate methods effectively assume that the measured variables are all continuous. Indeed, many methods further assume that the data can be approximated by the multivariate normal distribution, possibly after they have been transformed. But in practice some of the variables may be discrete and may require special procedures to handle them.

Discrete data arise in various forms. Grouped data arise when continuous measurements are assigned to a discrete set of categories such as 'high', 'medium' and 'low'. But many variables are inherently discrete. A discrete variable is called *ordinal* if its possible categories can be arranged in some numerical order. A discrete variable is called *nominal* if its possible categories are designated by names rather than numbers. The possible categories for a discrete variable should be mutually exclusive and exhaustive. A variable which can only take two possible values is called a *binary* or *dichotomous* variable. As an example, consider a sample of voters which are classified by sex (male or female), by social class (A, B, C1, C2, D or E), and by marital status (single, married, divorced, separated). Here 'sex' is a binary variable, 'social class' is an ordinal variable, while 'marital status' is a nominal variable which is qualitative in that there is no numerical way of ordering the possible categories.

In this section we briefly consider the case where *all* the variables are discrete. From the data the analyst can count the number of observations in each combination of categories. With two variables, one then gets the familiar two-way *contingency* table of frequencies, while with more than two variables, we get what is called a multidimensional contingency table. Multivariate discrete data in this sort of form are variously called *categorical* data, *frequency* data, *count* data or *attribute* data.

The simplest type of two-way contingency table has just two rows and two columns, and is usually considered separately for a variety of reasons (see Everitt, 1977, Chapter 2). The more general two-way table has r rows and c columns and is called an $(r \times c)$ table. The analysis of such tables is discussed by Everitt (1977, Chapter 3). The reader is probably familiar with the χ^2 test for testing whether or not the rows and columns of a two-way table are independent. As a complement to such a test, we can calculate some measure of the association between the two variables, and various possible coefficients are discussed by Bishop *et al.* (1975, Chapter 11).

With more than two categorical variables, we have a multidimensional contingency table and the analysis of such tables is discussed by Everitt (1977, Chapter 4). Instead of just being interested in the single hypothesis, namely whether the variables are mutually independent, there may be several other hypotheses of interest, such as whether one particular variable is independent of the remainder.

In fact, attention has recently shifted away from measures of association and tests of hypotheses and on to model-building. Now in the general linear model the expected value of the response variable is additive in the effects of interest. But in a contingency table it is natural to look at the ratios of frequencies rather than at differences, which suggests that the effects of interest are multiplicative in their effect. This in turn suggests that one needs to take some sort of logarithmic transformation in order to get an additive model. *Log–linear models* (e.g., Everitt, 1977, Chapter 5) provide a systematic approach to describing multidimensional contingency tables and to producing estimates of the effects of interest. In these models the logarithm of the expected cell frequency is assumed to be a linear function of the model parameters which may include row effects, column effects and interaction terms.

Another class of models is the *logistic* or *logit* model, which is particularly useful for explaining the variation in one binary response variable in terms of one or more explanatory variables which may be discrete or continuous (see Cox, 1970; Bishop *et al.*, 1975, Section 10.4). For example, in biological assay work we might measure the proportion of animals which die when injected with different doses of a poison. Here we are essentially interested in the probability p of 'success', but probabilities are confined to the range $[0, 1]$ and cannot be described directly by a linear model. The logistic transformation of p is given by $\log[p/(1-p)]$ and this quantity can take any value in the range $[-\infty, +\infty]$. It is often called the *logit* of p. The logistic model assumes that the logit of a cell probability is a linear function of the model parameters, and so the model is analogous to ordinary regression models. For a contingency table with two columns and fixed row totals, the logit formulation is equivalent to that of the log–linear model.

We have already remarked in Section 1.7 that the general linear model is outside the scope of this book, and this remark also applies to log–linear and logistic models. Several books have been entirely devoted to the analysis of discrete data (e.g., Plackett, 1974; Bishop *et al.*, 1975; Everitt, 1977).

Exercises

3.1 Show that the sample mean vector $\bar{\mathbf{x}}$, is an unbiased estimate of the population mean vector $\boldsymbol{\mu}$. Also show that the sample covariance matrix S is an unbiased estimate of the population covariance matrix Σ.

3.2 Show that the sample covariance matrix S is positive semidefinite. Hence show that the sample correlation matrix R is positive semidefinite.

[*Hint:* For rth individual, let $y_r = \sum_{j=1}^{p} a_j x_{rj}$, where $\mathbf{a}^T = [a_1, \ldots, a_p]$ is a vector of constants. Use the fact that $\sum_{r=1}^{n} (y_r - \bar{y})^2 \geq 0$ for every $\mathbf{a}$, and express this sum of squares in terms of S. Use Equation (3.6) for the second part.]

PART TWO
Finding new underlying variables

The next two chapters describe two variable-directed techniques, called principal component analysis and factor analysis. In both cases the main aim of the analysis is to replace the original variables by a smaller number of 'underlying' variables.

Principal component analysis consists of finding an orthogonal transformation of the original variables to a new set of uncorrelated variables, called principal components, which are derived in decreasing order of importance. These components are linear combinations of the original variables. The analyst often hopes that the first few components will account for most of the variation in the original data so that the effective dimensionality of the data can be reduced.

Factor analysis has somewhat similar aims but is based on a proper statistical model which specifies a given number of underlying variables called factors. The analysis is more concerned with 'explaining' the covariance structure of the variables rather than with 'explaining' the variances. Because of the similarities in the two approaches, the two methods are sometimes confused. However, there are fundamental differences in the underlying concepts.

In many reported applications, the principal components or factors appear to be an end in themselves, and researchers try to interpret them in a meaningful way. One common procedure is to split the variables into groups which are associated with particular components or factors. These groups often have the property that variables within the same group are highly correlated, while variables in different groups have low correlations. Frequently these groups can be found more easily by looking directly at the correlation matrix and grouping the variables in a common-sense way. We will argue that the main benefit of principal component analysis lies in allowing us to reduce the dimensionality of the problem so as to simplify subsequent analyses.

The use of factor analysis is particularly controversial for a variety of reasons, as outlined in Chapter 5. The method appears to be used very little by statisticians, though it is widely used (and misused) in the social sciences. Thus the statistician needs to know about the method if only to advise against its use in inappropriate situations.

CHAPTER FOUR

Principal component analysis

4.1 Introduction

In order to examine the relationships among a set of p correlated variables, it may be useful to transform the original set of variables to a new set of uncorrelated variables called *principal components*. These new variables are *linear* combinations of the original variables and are derived in decreasing order of importance so that, for example, the first principal component accounts for as much as possible of the variation in the original data. The transformation is in fact an *orthogonal rotation* in p-space.

The technique for finding this transformation is called principal component analysis (abbreviated to PCA). It is a *variable-directed* technique which is appropriate when the variables arise 'on an equal footing', so that, for example, we do *not* have a dependent variable and several explanatory variables as in multiple regression. PCA originated in some work by Karl Pearson around the turn of the century, and was further developed in the 1930s by Harold Hotelling and other workers using the approach described in Section 4.2.

The usual objective of the analysis is to see if the first few components account for most of the variation in the original data. If they do, then it is argued that the effective dimensionality of the problem is less than p. In other words, if some of the original variables are highly correlated, they are effectively 'saying the same thing' and there may be near-linear constraints on the variables. In this case it is hoped that the first few components will be intuitively meaningful, will help us understand the data better, and will be useful in subsequent analyses where we can operate with a smaller number of variables. In practice it is not always easy to give 'labels' to the components, and so we believe that their main use lies in reducing the dimensionality of the data in order to simplify later analyses. For example, plotting the scores of the first two components for each individual is a useful way of trying to find 'clusters' in the data (see Section 11.2) where one effectively reduces the dimensionality to two.

PCA transforms a set of correlated variables to a new set of uncorrelated variables. It is therefore worth stressing that, if the original variables are nearly uncorrelated, then there is no point in carrying out a PCA. The PCA will simply find components which are close to the original variables but arranged in decreasing order of variance (see Section 4.3.9).

Finally, we note that PCA is a *mathematical* technique which does not require the user to specify an underlying statistical model to explain the 'error' structure. In particular, no assumption is made about the probability distribution of the original variables, though more meaning can generally be given to the components in the case where the observations are assumed to be multivariate normal.

4.2 Derivation of principal components

Suppose $\mathbf{X}^T = [X_1, \ldots, X_p]$ is a p-dimensional random variable with mean $\boldsymbol{\mu}$ and covariance matrix Σ. Our problem is to find a new set of variables, say $Y_1, Y_2, \ldots, Y_p$, which are uncorrelated and whose variances decrease from first to last. Each Y_j is taken to be a linear combination of the X's, so that

$$\begin{aligned} Y_j &= a_{1j}X_1 + a_{2j}X_2 + \ldots + a_{pj}X_p \\ &= \mathbf{a}_j^T \mathbf{X} \end{aligned} \tag{4.1}$$

where $\mathbf{a}_j^T = [a_{1j}, \ldots, a_{pj}]$ is a vector of constants. Equation (4.1) contains an arbitrary scale factor. We therefore impose the condition that $\mathbf{a}_j^T\mathbf{a}_j = \sum_{k=1}^{p} a_{kj}^2 = 1$. We shall see that this particular normalization procedure ensures that the overall transformation is *orthogonal*—in other words, that distances in p-space are preserved.

The first principal component, Y_1, is found by choosing $\mathbf{a}_1$ so that Y_1 has the largest possible variance. In other words, we choose $\mathbf{a}_1$ so as to maximize the variance of $\mathbf{a}_1^T\mathbf{X}$ subject to the constraint that $\mathbf{a}_1^T\mathbf{a}_1 = 1$. This approach, originally suggested by Harold Hotelling, gives equivalent results to that of Karl Pearson, which finds the line in p-space such that the total sum of squared perpendicular distances from the points to the line is minimized.

The second principal component is found by choosing $\mathbf{a}_2$ so that Y_2 has the largest possible variance for all combinations of the form of Equation (4.1) which are uncorrelated with Y_1. Similarly, we derive $Y_3, \ldots, Y_p$, so as to be uncorrelated and to have decreasing variance.

We begin by finding the first component. We want to choose $\mathbf{a}_1$ so as to maximize the variance of Y_1 subject to the normalization constraint that $\mathbf{a}_1^T\mathbf{a}_1 = 1$. Now

$$\begin{aligned} \mathrm{Var}(Y_1) &= \mathrm{Var}(\mathbf{a}_1^T\mathbf{X}) \\ &= \mathbf{a}_1^T \Sigma \mathbf{a}_1 \end{aligned} \tag{4.2}$$

using Equation (2.8). Thus we take $\mathbf{a}_1^T \Sigma \mathbf{a}_1$ as our objective function.

The standard procedure for maximizing a function of several variables subject to one or more constraints is the method of Lagrange multipliers. With just one constraint, this method uses the fact that the stationary points of a differentiable function of p variables, say $f(x_1, \ldots, x_p)$, subject to a constraint $g(x_1, \ldots, x_p) = c$, are such that there exists a number λ, called the *Lagrange multiplier*, such that

$$\frac{\partial f}{\partial x_i} - \lambda \frac{\partial g}{\partial x_i} = 0 \qquad i = 1, \ldots, p \tag{4.3}$$

at the stationary points. These p equations, together with the constraint, are sufficient to determine the co-ordinates of the stationary points (and the corresponding values of λ, which, however, are usually of little interest). Further investigation is needed to see if a stationary point is a maximum, minimum or saddle point. It is helpful to form a new function, $L(\mathbf{x})$, such that

$$L(\mathbf{x}) = f(\mathbf{x}) - \lambda[g(\mathbf{x}) - c]$$

where the term in the square brackets is of course zero. Then the set of equations in (4.3) may be written simply as

$$\frac{\partial L}{\partial \mathbf{x}} = \mathbf{0}$$

using the definition given at the end of Section 1.6.

Applying this method to our problem, we write

$$L(\mathbf{a}_1) = \mathbf{a}_1^T \Sigma \, \mathbf{a}_1 - \lambda(\mathbf{a}_1^T \mathbf{a}_1 - 1)$$

Then, using Equation (1.18), we have

$$\frac{\partial L}{\partial \mathbf{a}_1} = 2\Sigma \, \mathbf{a}_1 - 2\lambda \mathbf{a}_1$$

Setting this equal to $\mathbf{0}$, we have

$$(\Sigma - \lambda I)\mathbf{a}_1 = \mathbf{0} \tag{4.4}$$

Note the insertion of the unit matrix I into Equation (4.4) so that the term in brackets is of the correct order, namely $(p \times p)$. We now come to the crucial step in the argument. If Equation (4.4) is to have a solution for $\mathbf{a}_1$, other than the null vector, then $(\Sigma - \lambda I)$ must be a singular matrix. Thus λ must be chosen so that

$$|\Sigma - \lambda I| = 0$$

Thus a non-zero solution for Equation (4.4) exists if and only if λ is an eigenvalue of Σ. But Σ will generally have p eigenvalues, which must all be nonnegative as Σ is positive semidefinite. Let us denote the eigenvalues by $\lambda_1, \lambda_2, \ldots, \lambda_p$, and assume for the moment that they are distinct, so that $\lambda_1 > \lambda_2 > \ldots > \lambda_p \geq 0$. Which one shall we choose to determine the first principal component? Now,

$$\begin{aligned} \operatorname{Var}(\mathbf{a}_1^T \mathbf{X}) &= \mathbf{a}_1^T \Sigma \, \mathbf{a}_1 \\ &= \mathbf{a}_1^T \lambda I \mathbf{a}_1 \qquad \text{using Equation (4.4)} \\ &= \lambda \end{aligned}$$

As we want to maximize this variance, we choose λ to be the *largest* eigenvalue, namely λ_1. Then, using Equation (4.4), the principal component, $\mathbf{a}_1$, which we are looking for must be the eigenvector of Σ corresponding to the largest eigenvalue.

The second principal component, namely $Y_2 = \mathbf{a}_2^T\mathbf{X}$, is obtained by an extension of the above argument. In addition to the scaling constraint that $\mathbf{a}_2^T\mathbf{a}_2 = 1$, we now have a second constraint that Y_2 should be uncorrelated with Y_1. Now,

$$\begin{aligned}\operatorname{Cov}(Y_2, Y_1) &= \operatorname{Cov}(\mathbf{a}_2^T\mathbf{X}, \mathbf{a}_1^T\mathbf{X}) \\ &= E[\mathbf{a}_2^T(\mathbf{X}-\mu)(\mathbf{X}-\mu)^T\mathbf{a}_1] \\ &= \mathbf{a}_2^T\Sigma\,\mathbf{a}_1 \qquad (4.5)\end{aligned}$$

We require this to be zero. But since $\Sigma\,\mathbf{a}_1 = \lambda_1\mathbf{a}_1$, an equivalent simpler condition is that $\mathbf{a}_2^T\mathbf{a}_1 = 0$. In other words, $\mathbf{a}_1$ and $\mathbf{a}_2$ should be orthogonal.

In order to maximize the variance of Y_2, namely $\mathbf{a}_2^T\Sigma\,\mathbf{a}_2$, subject to the two constraints, we need to introduce two Lagrange multipliers, which we will denote by λ and δ, and consider the function

$$L(\mathbf{a}_2) = \mathbf{a}_2^T\Sigma\,\mathbf{a}_2 - \lambda(\mathbf{a}_2^T\mathbf{a}_2 - 1) - \delta\mathbf{a}_2^T\mathbf{a}_1$$

At the stationary point(s) we must have

$$\frac{\partial L}{\partial \mathbf{a}_2} = 2(\Sigma - \lambda I)\mathbf{a}_2 - \delta\mathbf{a}_1 = \mathbf{0} \qquad (4.6)$$

If we premultiply this equation by $\mathbf{a}_1^T$, we obtain

$$2\mathbf{a}_1^T\Sigma\,\mathbf{a}_2 - \delta = 0$$

since $\mathbf{a}_1^T\mathbf{a}_2 = 0$. But from Equation (4.5), we also require $\mathbf{a}_1^T\Sigma\,\mathbf{a}_2$ to be zero, so that δ is zero at the stationary point(s). Thus Equation (4.6) becomes

$$(\Sigma - \lambda I)\mathbf{a}_2 = \mathbf{0}$$

With a little thought, we see that this time we choose λ to be the *second* largest eigenvalue of Σ, and $\mathbf{a}_2$ to be the corresponding eigenvector.

Continuing this argument, the jth principal component turns out to be the eigenvector associated with the jth largest eigenvalue.

There is no difficulty in extending the above argument to the case where some of the eigenvalues of Σ are equal. In this case there is no unique way of choosing the corresponding eigenvectors, but as long as the eigenvectors associated with multiple roots are chosen to be orthogonal, then the argument carries through.

Let us denote the $(p \times p)$ matrix of eigenvectors by A, where

$$A = [\mathbf{a}_1, \ldots, \mathbf{a}_p]$$

and the $(p \times 1)$ vector of principal components by $\mathbf{Y}$. Then

$$\mathbf{Y} = A^T\mathbf{X} \qquad (4.7)$$

The $(p \times p)$ covariance matrix of $\mathbf{Y}$ will be denoted by Λ and is clearly given by

$$\Lambda = \begin{bmatrix} \lambda_1 & 0 & \dots & 0 \\ 0 & \lambda_2 & \dots & 0 \\ \vdots & & & \\ 0 & & \dots & \lambda_p \end{bmatrix} \tag{4.8}$$

Note that the matrix is diagonal as the components have been chosen to be uncorrelated.

Using Equation (2.9), we can also express Var $(\mathbf{Y})$ in the form $A^T \Sigma A$, so that

$$\Lambda = A^T \Sigma A \tag{4.9}$$

gives the important relation between the covariance matrix of $\mathbf{X}$ and the corresponding principal components. Note that Equation (4.9) can be rewritten as

$$\Sigma = A \Lambda A^T \tag{4.10}$$

since A is an orthogonal matrix with $AA^T = I$.

We have already noted that the eigenvalues can be interpreted as the respective variances of the different components. Now the sum of these variances is given by

$$\sum_{i=1}^{p} \text{Var}(Y_i) = \sum_{i=1}^{p} \lambda_i = \text{trace}(\Lambda)$$

But

$$\begin{aligned} \text{trace}(\Lambda) &= \text{trace}(A^T \Sigma A) \\ &= \text{trace}(\Sigma A A^T) \qquad \text{using Equation (1.2)} \\ &= \text{trace}(\Sigma) \\ &= \sum_{i=1}^{p} \text{Var}(X_i) \end{aligned}$$

Thus we have the important result that the sums of the variances of the original variables and of their principal components are the same. It is therefore convenient to make statements such as 'the ith principal component accounts for a proportion $\lambda_i \Big/ \sum_{j=1}^{p} \lambda_j$ of the total variation in the original data', though it should be emphasized that this is not an analysis of variance in the usual sense of the expression. We will also say that the first m components account for a proportion $\sum_{j=1}^{m} \lambda_j \Big/ \sum_{j=1}^{p} \lambda_j$ of the total variation.

4.2.1 *Principal components from the correlation matrix*

It is quite common to calculate the principal components of a set of variables after they have been standardized to have unit variance. This means that one is effectively finding the principal components from the correlation matrix P rather than from the covariance matrix, Σ. The mathematical derivation is the same, so that the components turn out to be the eigenvectors of P. However, it is important to realize that the eigenvalues and eigenvectors of P will generally not be the same as those of Σ (see the discussion in Section 4.4). Choosing to analyse P rather than Σ involves a definite but arbitrary decision to make the variables 'equally important'.

For the correlation matrix, the diagonal terms are all unity. Thus the sum of the diagonal terms (or the sum of the variances of the standardized variables) will be equal to p. Thus the sum of the eigenvalues of P will also be equal to p, so that the proportion of the total variation accounted for by the jth component is simply λ_j/p.

4.2.2 *Estimating the principal components*

The above derivation of the principal components of $\mathbf{X}$ assumes that Σ is known. Generally this will not be so, and Σ is replaced by S, the sample covariance matrix, given by Equation (3.4). The derivation of the principal components of $\mathbf{X}$ using the *sample* variances and covariances is much as before. The principal components are found to be the eigenvectors of S. Let us denote the eigenvalues of S in descending order of size by $\hat{\lambda}_1, \hat{\lambda}_2, \ldots, \hat{\lambda}_p$, and the corresponding eigenvectors by $\hat{\mathbf{a}}_1, \ldots, \hat{\mathbf{a}}_p$. Since S is positive semidefinite, the eigenvalues are all non-negative and represent the estimated variances of the different components.

If our sample of 'individuals' is a random sample from a larger population, then $\{\hat{\lambda}_i\}$ and $\{\hat{\mathbf{a}}_i\}$ may be regarded as estimates of the eigenvalues and vectors of Σ, giving us estimates of the principal components of $\mathbf{X}$. But no assumptions have hitherto been made about the underlying population, and without such assumptions it is impossible to derive the sampling properties of the estimates. If we are prepared to assume that the observations are taken from a multivariate normal distribution, then some sampling theory is available (e.g., see Morrison, 1976, Section 8.7). But this theory is of limited practical value, partly because many of the results are for the asymptotic case (as $n \to \infty$), and partly because the normality assumption is often questionable. In any case, the 'sample' may be observations for a complete population. Thus the modern tendency is to view PCA as a mathematical technique with no underlying statistical model. The principal components obtained from the sample covariance matrix S are seen as *the* principal components and not as estimates of the corresponding quantities obtained from Σ. The 'hats' over λ_i and $\mathbf{a}_i$ are often omitted. Indeed, it is not even necessary to regard $\mathbf{X}$ and $\mathbf{Y}$ as random variables.

Example 4.1. Suppose we have just two standardized variables, X_1 and X_2, whose correlation matrix is

$$\mathrm{P} = \begin{bmatrix} 1 & \rho \\ \rho & 1 \end{bmatrix}$$

Find the principal components of $\mathbf{X}$, where $\mathbf{X}^{\mathrm{T}} = [X_1, X_2]$.

In order to find the principal components, we find the eigenvalues and eigenvectors of P. The eigenvalues are the roots of $|P - \lambda I| = 0$, which gives $(1-\lambda)^2 - \rho^2 = 0$. Thus the eigenvalues are $(1+\rho)$ and $(1-\rho)$. Note that the sum of the eigenvalues is equal to the sum of the diagonal terms of P, namely 2.

If $\rho > 0$, then the eigenvalues in order of magnitude are $\lambda_1 = (1+\rho)$ and $\lambda_2 = (1-\rho)$. The eigenvector corresponding to $\lambda_1 = (1+\rho)$, namely $\mathbf{a}_1^T = [a_{11}, a_{21}]$, is obtained by solving $\mathbf{P}\mathbf{a}_1 = \lambda_1 \mathbf{a}_1$, or the two equations $a_{11} + \rho a_{21} = (1+\rho)a_{11}$ and $\rho a_{11} + a_{21} = (1+\rho)a_{21}$. These two equations are of course identical and both reduce to the equation $a_{11} = a_{21}$, so that there is no unique solution. To obtain the principal components we introduce the normalization constraint that $\mathbf{a}_1^T\mathbf{a}_1 = a_{11}^2 + a_{21}^2 = 1$, and then we find $a_{11} = a_{21} = 1/\sqrt{2}$. Similarly, we find that the second eigenvector is given by $\mathbf{a}_2^T = [1/\sqrt{2}, -1/\sqrt{2}]$. Thus the principal components are given by

$$Y_1 = (X_1 + X_2)/\sqrt{2}$$

and

$$Y_2 = (X_1 - X_2)/\sqrt{2} \qquad (4.11)$$

Here Y_1 is the normalized average of the two variables, while Y_2 is the normalized difference.

If $\rho < 0$, the order of the eigenvalues, and hence of the principal components, will be reversed. If $\rho = 0$, the eigenvalues are both equal to 1 and any two components at right angles could be chosen, such as the two original variables. But as the variables are uncorrelated in this case, a PCA is pointless.

Three points to notice about Equation (4.11) are:

(*a*) There is an arbitrary sign in the choice of $\mathbf{a}_i$ and hence of Y_i. It is customary to choose a_{1i} to be positive.

(*b*) The components do not depend on ρ. This may be surprising at first sight, but should be obvious on second thoughts in view of the symmetry between the two variables. However, although the components stay the same, the proportion of variance explained by the first component, namely $(1+\rho)/2$, *does* change with ρ. In particular, as ρ approaches $+1$, the first component accounts for nearly all the variance, while as ρ approaches zero, the first and second components both account for half the total variance.

(*c*) In the bivariate case, the *sample* correlation matrix R, given by

$$R = \begin{bmatrix} 1 & r \\ r & 1 \end{bmatrix}$$

is of exactly the same form as P. So it follows that the principal components of R will be the same as those of P. This will generally not happen with more than two variables. □

4.3 Further results on PCA

This section gives a number of useful results regarding PCA.

4.3.1 *Mean-corrected component scores*

Equation (4.7) relates the observed vector random variable, $\mathbf{X}$, to the principal components, $\mathbf{Y}$, in such a way that $\mathbf{Y}$ will generally have a non-zero mean. It is more usual to add an appropriate vector of constants so that the principal components all have zero mean. If the sample mean is $\bar{\mathbf{x}}$, and A is the matrix of eigenvectors for the sample covariance matrix, S, then the usual transformation is

$$\mathbf{Y} = A^{\mathrm{T}}(\mathbf{X} - \bar{\mathbf{x}}) \tag{4.12}$$

This consists of a translation followed by an orthogonal rotation.

Using Equation (4.12) on the observation $\mathbf{x}_r$ for the rth individual, we find

$$\mathbf{y}_r = A^{\mathrm{T}}(\mathbf{x}_r - \bar{\mathbf{x}}) \tag{4.12a}$$

and the terms of $\mathbf{y}_r$ are called the *component scores* for the rth individual. It is essential to note that if A denotes the matrix of eigenvectors for the sample *correlation* matrix, then Equation (4.12a) should only be used after standardizing the observations, $(\mathbf{x}_r - \bar{\mathbf{x}})$, so that each variable has unit variance.

4.3.2 *The inverse transformation*

This is

$$\mathbf{X} = A\mathbf{Y} + \bar{\mathbf{x}} \tag{4.13}$$

Both Equations (4.12) and (4.13) are linear transformations. From Equation (4.12), the jth component depends on the coefficients in the jth row of A^{T} (or the jth column of A). The same overall set of coefficients appears in Equation (4.13), where the jth variable depends on the jth *row* of A. When studying an observed set of components, it is essential to know if one is looking at the matrix A, at its transpose, or at a related set of quantities called component correlations described below in Section 4.3.7.

4.3.3 *Zero eigenvalues*

If some of the original variables are linearly dependent, then some of the eigenvalues of Σ will be zero. The dimension of the space containing the observations is equal to the rank of Σ, and this is given by (p − number of zero eigenvalues). If there are k zero eigenvalues, then we can find k independent linear constraints on the variables. These constraints are sometimes called *structural* relationships (see Sprent, 1969, p. 32).

If we have a sample from a larger population, then an exact linear dependence in the population will also exist in the sample. If we look at the eigenvalues of S, then the sample estimates of the zero eigenvalues of Σ should also be zero (except perhaps for rounding error).

4.3.4 *Small eigenvalues*

The occurrence of exact linear dependence is rare, except where one deliberately introduces some redundant variables. A more important practical problem is to detect approximate linear dependence. If the smallest eigenvalue, λ_p, is very close to zero, then the pth principal component, $\mathbf{a}_p^T\mathbf{X}$, is 'almost' constant, and the dimension of $\mathbf{X}$ is 'almost' less than p. If the last few eigenvalues are judged to be small, then the 'efficiency' of restricting the dimension to, say, m is given by $\sum_{i=1}^{m} \lambda_i \Big/ \sum_{i=1}^{p} \lambda_i$. The principal components corresponding to small eigenvalues are variates for which the members of the population have almost equal values. These components can be taken as estimates of underlying linear relationships.

If $\lambda_{m+1}, \ldots, \lambda_p$ are 'small', little information is lost by replacing the values of the corresponding principal components with their means, which are chosen to be zero using Equation (4.12). Thus we can approximate the component scores for the rth individual by $\mathbf{y}_r^T = [y_{r1}, \ldots, y_{rm}, 0, \ldots, 0]$ and approximate the corresponding original observations by $\mathbf{x}_r = A\mathbf{y}_r + \bar{\mathbf{x}}$. We can also approximate S by $\sum_{i=1}^{m} \lambda_i \mathbf{a}_i \mathbf{a}_i^T$, this being left as an exercise for the reader (see Exercises 4.1 and 4.7). In both these formulae we use only the first m components and it is standard practice to look at only the first few columns of A corresponding to those eigenvalues which are thought to be 'large'.

4.3.5 *Repeated roots*

Sometimes some of the eigenvalues of Σ will be equal. If

$$\lambda_{q+1} = \ldots = \lambda_{q+k}$$

then $\lambda = \lambda_{q+1}$ is said to be a root of multiplicity k. The eigenvectors corresponding to multiple roots are not unique as one may choose any orthonormal set in the appropriate subspace of k dimensions. The corresponding principal components will have the same variance. The problem that arises in practice is that the corresponding roots of the *sample* covariance matrix will generally not be equal, so that the multiplicity is generally not observed in the sample case. Instead, we find that different samples will give completely different estimates of the eigenvectors and so they should not be regarded as 'characteristic variables'. This is something to bear in mind when eigenvalues are found to be nearly equal. Tests have been derived to test the equality of eigenvalues (e.g., Anderson, 1963) but these tests require a normality assumption and are valid only for large samples.

An important special case occurs when the last k eigenvalues are equal. In this case, the variation in the last k dimensions is said to be *spherical*. Then the last k principal components may be regarded as measuring non-specific variability and the essential characteristics of $\mathbf{X}$ are represented by the first $(p-k)$ principal components. Of course, if the last k eigenvalues are equal and zero, then the comments in Section 4.3.3 apply.

Example 4.2. Consider the $(p \times p)$ covariance matrix Σ whose diagonal terms are all unity and whose off-diagonal terms are all equal to ρ, where $0 < \rho < 1$.

The eigenvalues of Σ (see Exercise 1.5) are $\lambda_1 = 1 + (p-1)\rho$ and $\lambda_2 = \lambda_3 = \ldots = \lambda_p = (1-\rho)$.

Thus there are repeated roots if $p > 2$. In this special case, where only the first eigenvalue is distinct, we say that $\mathbf{X}$ is distributed spherically about a single principal axis. We leave it as an exercise for the reader to demonstrate that the eigenvector corresponding to λ_1 is of the simple form

$$\mathbf{a}_1^{\mathrm{T}} = [1/\sqrt{p}, \ldots, 1/\sqrt{p}]$$ □

4.3.6 *Orthogonality*

We have already noted that the matrix of eigenvectors A is an orthogonal matrix. The implication of using an orthogonal rotation is that the sum of squared deviations for each individual about the overall mean vector will be unchanged by the rotation. This can readily be demonstrated. Denoting the $(n \times p)$ mean-corrected data matrix by X and the $(n \times p)$ component-scores matrix by Y, we have

$$Y = XA \tag{4.14}$$

Thus

$$YY^{\mathrm{T}} = XAA^{\mathrm{T}}X^{\mathrm{T}} = XX^{\mathrm{T}} \tag{4.15}$$

The diagonal terms of the $(n \times n)$ matrices (YY^{T}) and (XX^{T}) give the sums of squares for each individual.

4.3.7 *Component loadings/component correlations*

When the principal components are tabulated, it is quite common to present the scaled vectors $\mathbf{a}_j^* = \lambda_j^{1/2}\mathbf{a}_j$, for $j = 1, 2, \ldots, p$, rather than the eigenvectors $\{\mathbf{a}_j\}$. These scaled vectors are such that the sum of squares of the elements is equal to the corresponding eigenvalue λ_j, rather than unity, since $\mathbf{a}_j^{*\mathrm{T}}\mathbf{a}_j^* = \lambda_j\mathbf{a}_j^{\mathrm{T}}\mathbf{a}_j = \lambda_j$.

Setting $C = [\mathbf{a}_1^*, \mathbf{a}_2^*, \ldots, \mathbf{a}_p^*]$, we see that $C = A\Lambda^{1/2}$ so that, from Equation (4.10), $\Sigma = CC^{\mathrm{T}}$. The elements of C are such that the coefficients of the more important components are scaled to be generally larger than those of the less important components, which seems intuitively sensible.

The scaled vectors $\{\mathbf{a}_j^*\}$ have two direct interpretations. Firstly, suppose we scale the components so that they all have unit variance. This can be done by $\mathbf{Y}^* = \Lambda^{-1/2}\mathbf{Y}$. Then the inverse transformation, $\mathbf{X} = A\mathbf{Y}$ (assuming $\mathbf{X}$ has mean zero) becomes $\mathbf{X} = A\Lambda^{1/2}\mathbf{Y}^* = C\mathbf{Y}^*$. Comparing this equation with the factor-analysis model to be given in Chapter 5, we see that the elements of C are analogous to the coefficients called factor loadings, and so might be called *component loadings*.

A second interpretation of C arises when the correlation matrix P of $\mathbf{X}$ has been analysed so that $P = CC^{\mathrm{T}}$, Now

$$\begin{aligned}\mathrm{Cov}(Y_j, X_i) &= \mathrm{Cov}\left(Y_j, \sum_{k=1}^{p} a_{ik}Y_k\right)\\ &= a_{ij}\mathrm{Var}(Y_j)\\ &= a_{ij}\lambda_j\end{aligned}$$

since $\text{Var}(Y_j) = \lambda_j$. Now the $\{X_i\}$ have been standardized to have unit variance, so that

$$\begin{aligned}\text{Correlation}(Y_j, X_i) &= \lambda_j a_{ij}/\lambda_j^{1/2}\\ &= a_{ij}\lambda_j^{1/2}\end{aligned}$$

Thus the matrix of correlations is given by

$$\text{Correlation}(\mathbf{Y}, \mathbf{X}) = \Lambda^{1/2} A^{\mathrm{T}} = C^{\mathrm{T}}$$

Thus, when C^{T} is calculated from a correlation matrix P, its elements measure the correlations between the components and the original (standardized) variables, and these elements are then called the *component correlations.*

4.3.8 *Off-diagonal structure*

An interesting feature of components is that they depend on the ratios of the correlations, and not on their absolute values. If we divide all the off-diagonal elements of a correlation matrix by a constant k, such that $k > 1$, then we show below that the eigenvalues change but that the eigenvectors, and hence the principal components, do not change. This result applies to both sample and population correlation matrices. This feature means that correlation matrices which look quite different in kind can have the same components, thus emphasizing that we need to look at the eigenvalues when trying to interpret components.

The result has already been demonstrated in Example 4.1 for the case $p = 2$. In the case $p > 2$, let R denote a correlation matrix whose (i, j)th term is denoted by r_{ij}, and let R^* denote a correlation matrix whose (i, j)th term for $i \neq j$) is r_{ij}/k, where k is a constant such that $k > 1$. Then we have

$$R^* = R/k + (k-1)I/k$$

The eigenvectors of R^* are solutions of

$$\begin{aligned}&(R^* - \lambda^* I)\mathbf{a}^* = \mathbf{0}\\ \text{or } &(R/k + (k-1)I/k - \lambda^* I)\mathbf{a}^* = \mathbf{0}\\ \text{or } &(R - (k\lambda^* - k + 1)I)\mathbf{a}^* = \mathbf{0}\end{aligned}$$

But the eigenvectors of R are solutions of

$$(R - \lambda I)\mathbf{a} = \mathbf{0}$$

So clearly the eigenvectors are the same, and the eigenvalues are related by

$$k\lambda^* - k + 1 = \lambda \qquad \text{or} \qquad \lambda^* = (\lambda + k - 1)/k$$

As $k \to \infty$, all the eigenvalues tend to unity. This is the result one would expect because, as $k \to \infty$, the off-diagonal terms of R^* tend to zero and we are left with p uncorrelated standardized variables. In this case, the components will each account for the same proportion $1/p$ of the total 'variance', which is p for a $(p \times p)$ correlation matrix.

4.3.9 *Uncorrelated variables*

Suppose one variable, say X_i, is uncorrelated with all other variables and has variance λ_i. Then it is easy to show that λ_i is an eigenvalue of the covariance matrix with the corresponding eigenvector having a 'one' in the ith place and zeroes elsewhere. Thus X_i is itself a principal component.

More generally, if a set of variables is mutually uncorrelated, then it is easy to show that the principal components are the same as the original variables but arranged in decreasing order of variance. In this situation, a PCA is of no assistance as there is no way of reducing the dimensionality of the problem.

As a cautionary tale, we recall the sociologist who came to us for help in interpreting the first three components of a (30×30) correlation matrix. These three components only accounted for 15 per cent of the total (standardized) variance. Now, even if the variables were all completely uncorrelated so that the eigenvalues were all unity, the first three components would still account for $100 \times 3/30 = 10$ per cent of the total variance. We therefore judged that the 'components' were not real features of the data so that there was no point in trying to interpret them. Looking back at the correlation matrix, we confirmed our suspicion that most of the correlations were 'small' so that the sociologist was wasting his time in carrying out a PCA.

Example 4.1 (*continued*). In example 4.1 we carried out a PCA of two variables, X_1 and X_2, whose correlation is equal to ρ. Suppose we extend this problem by adding a third standardized variable, X_3, which is uncorrelated with X_1 and X_2 so that the (3×3) correlation matrix is given by

$$\mathrm{P} = \begin{bmatrix} 1 & \rho & 0 \\ \rho & 1 & 0 \\ 0 & 0 & 1 \end{bmatrix}$$

What now are the principal components of $\mathbf{X}$, where

$$\mathbf{X}^{\mathrm{T}} = [X_1, X_2, X_3]?$$

The eigenvalues of P are found to be $(1 + \rho)$, $(1 - \rho)$ and 1. The normalized eigenvector corresponding to the unity eigenvalue is $[0, 0, 1]$, so that the corresponding principal component is simply the variable X_3. The other two principal components are the same linear combinations of X_1 and X_2 as obtained previously in the two-dimensional case. This illustrates the general point that the principal components of a set of variables are not affected by the addition of further uncorrelated variables, though they will of course be affected by the addition of correlated variables. □

4.4 The problem of scaling in PCA

It is important to realize that the principal components of a set of variables depend critically upon the scales used to measure the variables. For example, suppose that for each of n individuals we measure their weight in pounds, their height in feet and their age in years to give a vector $\mathbf{x}$. Denote the resulting sample covariance matrix by S_x, its eigenvalues by $\{\lambda_i\}$ and its eigenvectors by $\{\mathbf{a}_i\}$. If we transform to new co-ordinates so that

$$\mathbf{z}^{\mathrm{T}} = (\text{weight in kilograms, height in metres, age in months})$$

then $\mathbf{z} = K\mathbf{x}$, where K is the diagonal matrix

$$K = \begin{bmatrix} 1/2.2 & 0 & 0 \\ 0 & 1/3.28 & 0 \\ 0 & 0 & 12 \end{bmatrix}$$

as there are, for example, 2.2 pounds in one kilogram. The covariance matrix of the new variables will be given by

$$S_z = KS_xK$$

since $K^{\mathrm{T}} = K$. The eigenvalues and eigenvectors of S_z will generally be different from those of S_x and will be denoted by $\{\lambda^*_i\}$ and $\{\mathbf{a}^*_i\}$. But will they give the same principal components when transformed back to the original variables? The answer generally is no. We show below that the principal components will change unless:

(*a*) All the diagonal elements of K are the same, so that $K = cI$, where c is a scalar constant. This would mean that all the variables are scaled in the same way

(*b*) Variables corresponding to unequal diagonal elements of K are uncorrelated. In particular, if all the elements of K are unequal, then S_x must be a diagonal matrix, in which case there is no point in carrying out a PCA as the variables are all uncorrelated.

Proof (*This proof may be omitted at a first reading.*)
Consider the first PC of S_x, for which

$$S_x\mathbf{a}_1 = \lambda_1\mathbf{a}_1 \tag{4.16}$$

The first PC of S_z is given by

$$S_z\mathbf{a}^*_1 = \lambda^*_1\mathbf{a}^*_1$$

or

$$KS_xK\mathbf{a}^*_1 = \lambda^*_1\mathbf{a}^*_1 \tag{4.17}$$

Multiplying Equation (4.16) by K gives

$$KS_x\mathbf{a}_1 = \lambda_1K\mathbf{a}_1 \tag{4.18}$$

Now, if the first PC of S_x is the same as the first PC of S_z when written in terms of the original co-ordinates, then $y^*_1 = \mathbf{a}^{*\mathrm{T}}_1\mathbf{z} = \mathbf{a}^{*\mathrm{T}}_1K\mathbf{x}$ must be proportional to $y_1 = \mathbf{a}^{\mathrm{T}}_1\mathbf{x}$ for every $\mathbf{x}$, so that

$$\mathbf{a}^{*\mathrm{T}}_1K = c\mathbf{a}^{\mathrm{T}}_1$$

where c denotes a constant, or

$$\mathbf{a}_1^* = cK^{-1}\mathbf{a}_1 \tag{4.19}$$

Note that here we assume there are no zero elements on the diagonal of K so that K^{-1} exists. Another way of looking at Equation (4.19) is to note that $\mathbf{a}_1$ and $K\mathbf{a}_1^*$ are the same when the latter is normalized.

Substituting Equation (4.19) into Equation (4.17), we get

$$KS_x\mathbf{a}_1 = \lambda_1^* K^{-1}\mathbf{a}_1$$

Comparing with Equation (4.18), we see that the principal components are the same only if

$$\lambda_1 K\mathbf{a}_1 = \lambda_1^* K^{-1}\mathbf{a}_1$$

or

$$(\lambda_1 K^2 - \lambda_1^* I)\mathbf{a}_1 = \mathbf{0} \tag{4.20}$$

A similar result can be obtained for the other components. The equations can be satisfied if $(\lambda_i K^2 - \lambda_i^* I)$ is the null matrix or is singular.

If $\lambda_i K^2 = \lambda_i^* I$ for every i, then K^2 and hence K must be proportional to I. In this case, the eigenvalues will all increase in the same ratio but the eigenvectors of S_x and S_z will be the same, and so will the proportion of the total variance explained by each component.

Alternatively, if $(\lambda_i K^2 - \lambda_i^* I)$ is singular, then $|\lambda_i K^2 - \lambda_i^* I| = 0$ and λ_i^*/λ_i must be an eigenvalue of K^2, with $\mathbf{a}_i$ the corresponding eigenvector. But K^2 is diagonal so that its eigenvalues are the diagonal terms. Now if all the diagonal elements of K (and hence of K^2) are unequal, the eigenvectors of K^2 must be the elementary vectors; and these can only be the eigenvectors of S_x if S_x is diagonal, in which case the variables are uncorrelated. Alternatively, if some of the diagonal elements of K are equal, then the corresponding eigenvalues of K^2 will be equal and the eigenvectors corresponding to this multiple root can be chosen as any orthonormal set in the appropriate subspace. From Equation (4.20), we see that it must be possible to choose these eigenvectors so that they are also eigenvectors of S_x, and with careful thought it can be seen that the set of variables corresponding to equal elements in K may be correlated with each other but must be uncorrelated with all other variables. □

Discussion. The practical outcome of the above result is that principal components are generally changed by scaling and that they are therefore not a unique characteristic of the data. If, for example, one variable has a much larger variance than all the other variables, then this variable will dominate the first principal component of the covariance matrix whatever the correlation structure, whereas if the variables are all scaled to have unit variance, then the first principal component will be quite different in kind. Because of this, there is generally thought to be little point in carrying out a PCA unless the variables have 'roughly similar' variances, as may be the case, for example, if all the variables are percentages, or are measured in the same co-ordinates.

The conventional way of getting round the scaling problem is to analyse the *correlation* matrix rather than the covariance matrix, so that each multivariate observation, $\mathbf{x}$, is transformed by

$$\mathbf{z} = K\mathbf{x}$$

where

$$K = \begin{bmatrix} 1/s_1 & 0 & \dots & 0 \\ 0 & 1/s_2 & \dots & 0 \\ \vdots & & & \\ 0 & \dots & 0 & 1/s_p \end{bmatrix}$$

and s_i is the sample standard deviation for the ith variable. This ensures that all variables are scaled to have unit variance and so in some sense have equal importance. This scaling procedure is still arbitrary to some extent, is data-dependent and avoids rather than solves the scaling problem. If the variables are not thought to be of equal importance, then the analysis of the correlation matrix is not recommended. Analysing the correlation matrix also makes it more difficult to compare the results of PCA on two or more different samples.

Note that the principal components of the correlation matrix will not be orthogonal if the variables are transformed back to their original co-ordinates. This is because a linear transformation of two lines at right angles in P-space will not generally give two new lines at right angles, which is another way of explaining why the scaling problem arises in the first place.

In conclusion, we note that the scaling problem does not arise in correlation and regression. Correlation coefficients do not depend on the units in which the variables are measured, while regression equations are equivalent whatever scales are used (see Exercise 4.8). The latter result arises because regression equations are chosen to minimize the sum of squared distances parallel to one of the co-ordinate axes, while in PCA we have already noted that the first principal component is chosen so as to minimize the sum of squared distances perpendicular to a line in p-space.

4.5 Discussion

This section discusses the various objectives of PCA and assesses the benefits and drawbacks of applying the method in practice. With computer programs readily available, it is now very easy (perhaps too easy) to perform a PCA. But the subsequent interpretation and use of the components is by no means easy.

Multivariate analysts often say that the technique is exploratory, and should be used to get a 'feel' for a set of data. Hopefully, the method will lead the user to a better understanding of the correlation structure and may generate hypotheses regarding the relationships between the variables.

The two main objectives appear to be the identification of new meaningful underlying variables, and the reduction of the dimensionality of the problem as a prelude to further analysis of the data. A third related objective, namely the elimination of those original variables which contribute relatively little extra information, will not be considered here (see Joliffe, 1973).

4.5.1 *The identification of important components*

After calculating the eigenvalues and principal components of a correlation (or covariance) matrix, the usual procedure is to look at the first few components which, hopefully, account for a large proportion of the total variance. In order to do this, it is necessary to decide which eigenvalues are 'large' and which are 'small' so that components corresponding to the latter may be disregarded. When analysing a correlation matrix where the sum of the eigenvalues is p, many social scientists use the rule that eigenvalues less than 1 may be disregarded. This arbitrary policy is a useful rule of thumb but has no theoretical justification. It may be better to look at the pattern of eigenvalues and see if there is a natural breakpoint. The fact that there is no objective way of deciding how many components to retain is a serious drawback to the method.

Many researchers try to identify the components corresponding to 'large' eigenvalues as describing some underlying feature of the population. But this is often rather difficult.

One common type of situation arises when all the variables are positively correlated. The first principal component is then some sort of weighted average of the variables and can be regarded as a measure of *size*. In particular, if the correlation matrix is analysed, the (standardized) variables will have nearly equal weights. For example, the correlation matrix

$$\begin{bmatrix} 1 & 0.4 & 0.4 & 0.4 \\ & 1 & 0.4 & 0.4 \\ & & 1 & 0.4 \\ & & & 1 \end{bmatrix}$$

has $\lambda_1 = 2.2$ and $\mathbf{a}_1^{\mathrm{T}} = [0.5, 0.5, 0.5, 0.5]$. Here the weights happen to be exactly equal.

The correlation matrix

$$\begin{bmatrix} 1 & 0.9 & 0.3 & 0.3 \\ & 1 & 0.4 & 0.4 \\ & & 1 & 0.9 \\ & & & 1 \end{bmatrix}$$

may appear fairly dissimilar to that given above, but its first principal component, namely $\mathbf{a}_1^{\mathrm{T}} = [0.48, 0.52, 0.50, 0.50]$, is nearly identical. This is typical for correlation matrices with positive elements, suggesting that there is little point in looking at $\mathbf{a}_1$ in this case. It might be expected that the second principal component will then be of more interest, but in the above two examples we find $\lambda_2 < 1$ in both cases, indicating that there is not much point in looking at $\mathbf{a}_2$ either.

For correlation matrices containing both positive and negative elements, the position is less clear. In trying to attach a meaningful label to a particular component, the usual procedure seems to consist of looking at the corresponding eigenvector and picking out the variables for which the coefficients in the eigenvector are relatively large, either positive or negative. Having established the subset of variables which is important for a particular component, the user then tries to see what these variables have in common. Sometimes the first few components are *rotated* in order to find a new set of components which can more easily be interpreted. The rotation is usually, though not necessarily, orthogonal. When using rotation, the usual procedure is to carry out a PCA, subjectively choose the number of important components – say m, and then calculate linear combinations of the selected eigenvectors in the subspace of m dimensions so as to get a new set of components (which will no longer be 'principal') satisfying some desired property. For example, given the truncated version of Equation (4.13) with the matrix A of order $(p \times m)$, the *varimax* method tries to find a new A where the coefficients are relatively large or relatively small compared with the original ones. The idea is that each variable should be heavily loaded on as few components as possible. This rotation procedure is sometimes incorrectly called *factor analysis*. This latter term should strictly be reserved for the procedures described in Chapter 5, which depend on a proper statistical model.

In trying to identify components in this sort of way, the PCA often seems to be an end in itself and a few case studies are available (see Kendall, 1975, Chapter 2; Jeffers, 1967). But if the components can be used to group the variables, then it is often the case that these groups can be found by direct visual inspection of the correlation matrix, such that variables within a group are highly correlated while variables in different groups have 'low' correlation (see also Sections 3.2.1 and 11.1.2; Kamen, 1970; Kendall, 1975, Example 3.1; Chakrapani and Ehrenberg, 1979). But some authors (e.g., Jeffers, 1967, p. 229) argue that it is quicker and easier to extract the eigenvalues and eigenvectors of a correlation matrix than to spend time looking at the matrix. There may be some truth in the latter view when there are a large number of variables, but it is also fair to point out that various workers have reported trying PCA without being able to give the resulting components any practical interpretation. It is also relevant to remind the reader that there is no underlying statistical model in PCA and that the scaling problem is a further drawback. Thus we conclude that it is rather dangerous to try and read too much meaning into components (see also Kendall, 1975, p. 23).

Example 4.3. In order to illustrate the previous remarks on identifying and interpreting important components, we will reconsider the crime data given in Table 3.2. These data were analysed by Ahamad (1967) using PCA in order 'to investigate the relationships between different crimes and to determine to what extent the variation in the number of crimes from year to year may be explained by a small number of unrelated factors'. The value of carrying out a PCA on these data is open to argument, and so we shall use the data both for an illustrative example of the use of PCA as well as to comment on the general value and dangers of trying to interpret components.

The variances of the original variables vary enormously as can be seen in Table 3.2 by comparing, for example, X_1 (homicide) with X_7 (larceny). Thus a PCA of the covariance matrix would give components which, to a large extent, would be the original variables in descending order of variance. Thus Ahamad chose (without explanation) to analyse the (18×18) correlation matrix. First the eigenvalues were calculated. The six largest values are 12.90, 2.71, 0.96, 0.68, 0.32 and 0.17, but the remainder are very small. The sum of the eigenvalues must be 18, so the remaining eigenvalues sum to only $(18 - 17.74) = 0.26$. Thus the first six components account for over 98 per cent $(= 100 \times 17.74/18)$ of the total variation. In fact, most statisticians would judge the fourth, fifth and sixth eigenvalues also to be 'small', leaving only the first three components to consider, which together account for 92 per cent of the total variation.

The component correlations for the first two components are shown in Table 4.1. When trying to interpret components, it is usually helpful to omit 'small' correlations, say those less than 0.25 in absolute magnitude, as one is interested in seeing which coefficients are 'large', and this rule has been followed here.

How can these two components be interpreted? The first component is highly correlated with all the variables in the group which give high positive correlations with each other in Table 3.4. Thus, this component might be interpreted as measuring the general increase in crime due to changes in population structure, social conditions or whatever. Indeed, this is the interpretation made by Ahamad (1967). But there are various practical reasons

Table 4.1 Component correlations for the first two components of the data in Table 3.2

Variable	$\sqrt{(\lambda_1)}\mathbf{a}_1$	$\sqrt{(\lambda_2)}\mathbf{a}_2$
Homicide (X_1)		54
Woundings (X_2)	.97	
Homosexual offences (X_3)	−.31	.81
Heterosexual offences (X_4)	.92	−.29
Breaking and entering (X_5)	.99	
Robbery (X_6)	.98	
Larceny (X_7)	.99	
Frauds (X_8)	.98	
Receiving stolen goods (X_9)	.96	
Malicious injuries to property (X_{10})	.64	.57
Forgery (X_{11})	.97	
Blackmail (X_{12})	.96	
Assault (X_{13})	−.42	.73
Malicious damage (X_{14})	.94	
Revenue laws (X_{15})	.95	
Intoxication laws (X_{16})	.94	−.28
Indecent exposure (X_{17})	.34	.75
Taking motor vehicle without consent (X_{18})	.96	

(see Walker, 1967) why this component should *not* be regarded as a general 'index of crime'. In particular, by analysing the correlation matrix, the crime of larceny, with about 300 000 offences per year, is given the same weight as robbery, with only about 1000 offences per year.

The second component is much more difficult to identify. It is highly correlated with the remaining five of the original variables, namely X_1, X_3, X_{10}, X_{13} and X_{17}. These are the variables which do *not* show a steady increase over the given time period and we doubt if anything else can be said. The suggestion in Ahamad (1967) that this component may 'reflect changes in recording practice by the police over the period' seems rather dubious.

What about the third component, whose eigenvalue, 0.96, is neither obviously 'large' nor 'small'? Ahamad (1967) says that identification of the third component is 'extremely difficult'. We would go further and suggest that the third component is probably spurious. It is highly correlated only with homicide, and might well change completely if the data were collected over a slightly different time-period.

Indeed, the identification of all three components is to a considerable extent arbitrary in that completely different results would be produced by analysing the covariance matrix rather than the correlation matrix, or by analysing data from a somewhat different time-period. This suggests that it is dangerous to try to attach 'meaning' to the components in this case and that the value of PCA is open to question. □

The above example is rather atypical in that it involves time-series data, but it does suggest that PCA is not always the powerful technique that some people expect it to be. At best, it may give one ideas about how to group the variables. At worst, it may give components which appear to have no easy interpretation or which can be interpreted in a misleading way. So, in the next subsection, we go on to suggest that a more important objective in PCA lies in attempting to reduce the dimensionality of the data as a prelude to further analysis.

4.5.2 *The use of components in subsequent analyses*

An important benefit of PCA is that it provides a quick way of assessing the effective dimensionality of a set of data. If the first few components account for most of the variation in the original data, it is often a good idea to use these first few component scores in subsequent analyses. No distributional assumptions need to be made to do this, and one does not need to try to interpret the principal components *en route*.

As one example, we pointed out in Section 3.3 that it is always a good idea to plot data if possible. With more than three variables, this can be difficult. But if the first two components account for a large proportion of the total variation, then it will often be useful to plot the values of the first two component scores for each individual. In other words, PCA enables us to plot the data in two dimensions. In particular, one can then look for outliers or for groups or

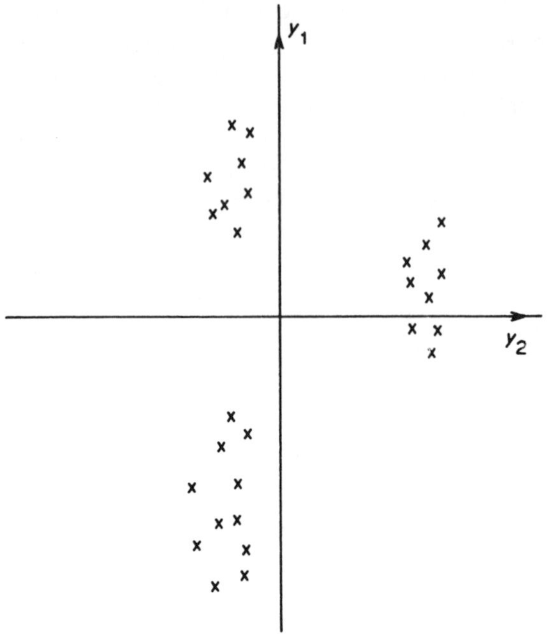

Figure 4.1 The values of the first two components for data collected on 26 individuals.

'clusters' of individuals, as further discussed in Section 11.2. An example where three groups of individuals can be seen in the graph of the first two components is shown in Fig. 4.1. The procedure of using component scores to produce a 'map' of the individuals is sometimes called *principal co-ordinates analysis* or *classical scaling* (see Section 10.3.2).

The dual of the above idea is to plot for each variable the coefficients of the first two components in the inverse transformation, Equation (4.13), and see if this reveals any natural grouping of the variables. In other words, we plot a_{i1} against a_{i2} for each i and look for clusters of points. This is an important use of PCA and often reveals groupings of variables which would not be found by other means.

As another example of the use of PCA in reducing dimensionality, one of the authors recently attempted a discriminant analysis (see Section 7.8) where the data consisted of two groups of 10 observations on 28 highly correlated variables. As the number of observations is less than the number of variables, there will be unpleasant singularity problems unless the dimensionality is drastically reduced. A PCA revealed two important dimensions and an effective discriminant function was constructed using these two new variables rather than the original 28.

As a third example, we note that multiple regression can be dangerous if the so-called independent variables are in fact highly correlated. Various techniques, such as ridge regression, have been developed to overcome this problem. An alternative approach, which may well be more fruitful, is to regress, not on the original variables, but on those components which are most highly correlated with the dependent variable (see Kendall, 1975, Section 7.23).

4.6 PCA for multivariate normal data

Hitherto, we have made no distributional assumptions about the data. In this section we investigate what happens when the observations are taken from a multivariate normal distribution, so that $\mathbf{X} \sim N_p(\boldsymbol{\mu}, \Sigma)$.

The mathematical derivation of the principal components of $\mathbf{X}$ is unchanged. If A denotes the matrix of eigenvectors of Σ, then the principal components may be written in the form

$$\mathbf{Y} = A^{\mathrm{T}}(\mathbf{X} - \boldsymbol{\mu}) \tag{4.21}$$

This is the same as Equation (4.7) except that, in addition to the orthogonal rotation, we have inserted a shift of origin to make $\mathbf{Y}$ have a zero mean vector. This is analogous to the mean-corrected formula (4.12) for the sample case.

The difference now is that, given the distribution of $\mathbf{X}$, we can find the distribution of $\mathbf{Y}$. Since each component of $\mathbf{Y}$ is a linear combination of normally distributed random variables, it follows that each component of $\mathbf{Y}$ will also be normally distributed. The reader should therefore not be too surprised to learn that the joint distribution of $\mathbf{Y}$ is multivariate normal, and this result is proved below. The mean vector of $\mathbf{Y}$ is $\mathbf{0}$ and its covariance matrix is given, as before, by Λ, which is the diagonal matrix whose ith diagonal element is λ_i. Thus in the notation of Section 2.3, we have to show that $\mathbf{Y} \sim N_p(\mathbf{0}, \Lambda)$.

As Σ must be non-singular, it follows that A is also non-singular, so that Equation (4.21) describes a one-to-one transformation. For one-to-one transformations, it can be shown that the joint density functions of $\mathbf{X}$ and $\mathbf{Y}$, say $f(\mathbf{x})$ and $g(\mathbf{y})$, are related by

$$g(\mathbf{y}) = f(\mathbf{x})_y |J| \tag{4.22}$$

where $|J|$ = Jacobian of the transformation

$$= \begin{vmatrix} \dfrac{\partial x_1}{\partial y_1} & \dfrac{\partial x_1}{\partial y_2} & \cdots & \dfrac{\partial x_1}{\partial y_p} \\ \vdots & & & \\ \dfrac{\partial x_p}{\partial y_1} & & & \dfrac{\partial x_p}{\partial y_p} \end{vmatrix}$$

By writing $\mathbf{X} = A\mathbf{Y} + \boldsymbol{\mu}$, we see that the Jacobian must be equal to $|A|$, which is equal to 1, since A is orthogonal with $AA^{\mathrm{T}} = I$ and $|A||A^{\mathrm{T}}| = 1$. Now in Equation (4.22), $f(\mathbf{x})_y$ means $f(\mathbf{x})$ evaluated at $\mathbf{y} = A^{\mathrm{T}}(\mathbf{x} - \boldsymbol{\mu})$. Since

$$f(\mathbf{x}) = \frac{1}{(2\pi)^{p/2}|\Sigma|^{1/2}} \exp\left[-\tfrac{1}{2}(\mathbf{x} - \boldsymbol{\mu})^{\mathrm{T}}\Sigma^{-1}(\mathbf{x} - \boldsymbol{\mu})\right]$$

we find

$$g(\mathbf{y}) = \frac{1}{(2\pi)^{p/2}|\Lambda|^{1/2}} \exp\left[-\tfrac{1}{2}\mathbf{y}^{\mathrm{T}}\Lambda^{-1}\mathbf{y}\right] \tag{4.23}$$

using $|\Sigma| = |\Lambda|$ and $\Sigma^{-1} = A\Lambda^{-1}A^{\mathrm{T}}$.

Writing Equation (4.23) in the form

$$g(\mathbf{y}) = \frac{1}{(2\pi)^{p/2}\left(\prod_{i=1}^{p} \lambda_i\right)^{1/2}} \exp\left[-\frac{1}{2}\sum_{i=1}^{p} y_i^2/\lambda_i\right]$$

we see that it is of the same form as the product of p independent normal density functions. Thus the result is proved.

In the multivariate normal case, the principal components have an obvious geometrical interpretation. The joint density function of $\mathbf{X}$ is constant on an ellipsoid in p dimensions. The principal components correspond to the *principal axes* of the ellipsoids. Indeed, the first principal component can be found in an alternative way by finding the chord of maximum distance inside the ellipsoid $(\mathbf{x} - \boldsymbol{\mu})^{\mathrm{T}}\Sigma^{-1}(\mathbf{x} - \boldsymbol{\mu}) = \text{constant}$.

It is of interest to consider the special case of bivariate normality where $p = 2$. The joint density function of $\mathbf{X}$ is constant on a set of ellipses (see Fig. 2.2). The principal components of $\mathbf{X}$ are such that $Y_1 = \text{constant}$ on the *minor* axis of these ellipses, while $Y_2 = \text{constant}$ on the *major* axes. Thus Y_1 varies most along the major axis.

Putting $\mathbf{Y} = A^{\mathrm{T}}(\mathbf{X} - \boldsymbol{\mu})$, the ellipse $(\mathbf{X} - \boldsymbol{\mu})^{\mathrm{T}}\Sigma^{-1}(\mathbf{X} - \boldsymbol{\mu}) = \text{constant}$ transforms to $\mathbf{Y}^{\mathrm{T}}\Lambda^{-1}\mathbf{Y} = \text{constant}$ (using $\Lambda = A^{\mathrm{T}}\Sigma A$), or

$$\frac{Y_1^2}{\lambda_1} + \frac{Y_2^2}{\lambda_2} = \text{constant}$$

Thus the effect of the transformation is to rotate the axes so that major and minor axes are parallel to the co-ordinate axes. If the eigenvalues of Σ are equal, the new 'ellipse' is a circle and there is no unique transformation. In fact, this only happens if the original variables are uncorrelated and have equal variance.

In the above discussion, we have assumed knowledge of $\boldsymbol{\mu}$ and Σ. In practice these parameters will be unknown and the observed values of $\mathbf{X}$ will constitute a sample from $N_p(\boldsymbol{\mu}, \Sigma)$. The corresponding component scores are calculated using the eigenvectors, not of Σ, but of the sample covariance matrix S, and so the component scores will only approximate a sample from $N_p(\mathbf{0}, \Lambda)$.

4.7 Summary

PCA consists of an orthogonal transformation in p-space from the original variables to a new set of variables, called principal components. The main stages in the analysis are:

(*a*) Decide if it is worth including all the variables recorded in the original data matrix, and whether any of the variables need to be transformed.
(*b*) Calculate the correlation (or covariance) matrix, bearing in mind that a correlation coefficient should generally not be calculated for a pair of variables whose relationship is obviously non-linear.
(*c*) Look at the correlation matrix and observe any natural groupings of variables with 'high' correlations. However, if nearly all the correlations are 'small', then there is probably not much point in carrying out a PCA.
(*d*) Calculate the eigenvalues and eigenvectors of the correlation (or covariance) matrix.
(*e*) Examine the eigenvalues and try to decide how many are 'large'. This should indicate the effective dimensionality of the data.
(*f*) Look at the groupings of variables suggested by the components and consider whether the components have some meaningful interpretation.
(*g*) Use the component scores in subsequent analyses as a way of reducing the dimensionality of the problem.

In Section 4.5, we argued that the latter application is probably the most useful rôle for principal components. But we must also recall the following drawbacks to PCA:

(i) It may be difficult or dangerous to try to read too much 'meaning' into the components.
(ii) The components are not invariant under linear transformations of the variables.
(iii) There is no underlying statistical model, and no provision can be made for variance components due to 'error'. This means that the sampling behaviour of the eigenvalues and eigenvectors is unknown. One result is that there is no objective way of deciding how many eigenvalues are 'large'. Another consequence is that it is difficult to compare the different components which result from carrying out a PCA on two or more different samples of the same type.

Exercises

4.1 Suppose the p-variate random variables $\mathbf{X}$ has covariance matrix Σ with eigenvalues $\{\lambda_i\}$ and orthonormal eigenvectors $\{\mathbf{a}_i\}$. Show that the identity matrix I is given by

$$I = \mathbf{a}_1 \mathbf{a}_1^{\mathrm{T}} + \ldots + \mathbf{a}_p \mathbf{a}_p^{\mathrm{T}}$$

and that

$$\Sigma = \lambda_1 \mathbf{a}_1 \mathbf{a}_1^{\mathrm{T}} + \ldots + \lambda_p \mathbf{a}_p \mathbf{a}_p^{\mathrm{T}}$$

The latter result is called the *spectral decomposition* of Σ, (see Equation 1.17). A similar result holds for the sample covariance matrix S.

4.2 Let X_1, X_2 be two random variables with covariance matrix

$$\Sigma = \begin{bmatrix} 9 & \sqrt{6} \\ \sqrt{6} & 4 \end{bmatrix}$$

Obtain the principal components, and find the percentage of the total variance explained by each component. Repeat the exercise using the corresponding correlation matrix P, and show that the principal components of P do not correspond to those of Σ. [*Hint*: In the last part of the question, remember to compare the components of Σ and P in the same units.]

4.3 The scores on two tests for each of 8 individuals were as follows:

Individual	Test 1	Test 2	Individual	Test 1	Test 2
1	2	9	5	10	13
2	16	20	6	4	4
3	8	0	7	10	7
4	18	11	8	12	16

Show that the sample covariance matrix is

$$\frac{1}{7}\begin{bmatrix} 208 & 144 \\ 144 & 292 \end{bmatrix}$$

and that the eigenvectors are [0.6, 0.8] and [0.8, − 0.6]. Compute the values of the mean-corrected principal components for each individual and check that the sum of products of the component scores is zero (because the components are uncorrelated).

Also find the sample correlation matrix and write down its principal components.

4.4 Suppose **X** is a 4-dimensional random variable with components X_1, X_2, X_3, X_4, which all have unit variance and that the covariance (and also correlation) matrix is given by

$$\Sigma = \begin{bmatrix} 1 & \frac{1}{2} & \frac{1}{2} & \frac{1}{2} \\ \frac{1}{2} & 1 & \frac{1}{2} & \frac{1}{2} \\ \frac{1}{2} & \frac{1}{2} & 1 & \frac{1}{2} \\ \frac{1}{2} & \frac{1}{2} & \frac{1}{2} & 1 \end{bmatrix}$$

Find the eigenvalues of Σ and show that the first principal component accounts for 62.5 per cent of the total variance of the X's. Write down the first principal component.

Suppose now that the scale of X_1 is doubled. Find the covariance matrix of $\mathbf{X}^*$, where $\mathbf{X}^{*T} = [2X_1, X_2, X_3, X_4]$. Show that the first principal component of $\mathbf{X}^*$ explains 5/7 of the total variance of $\mathbf{X}^*$. Find the first principal component and show that it is not invariant under change of scale.

4.5 Suppose $\mathbf{X}$ has covariance matrix

$$\Sigma = \begin{bmatrix} 3 & 1 & 1 \\ 1 & 3 & 1 \\ 1 & 1 & 5 \end{bmatrix}$$

Find the principal components of $\mathbf{X}$, and say what proportion of the total variance is accounted for by the first two principal components.

If $\mathbf{X}$ follows a multivariate normal distribution, what is the joint distribution of the (mean-corrected) principal components of $\mathbf{X}$?

4.6 Suppose (X_1, X_2) follow a bivariate normal distribution with zero means, unit variances and correlation ρ. By finding the eigenvectors of the covariance matrix, find the transformation which produces independent random variables.

4.7 Four measurements were made for each of a random sample of 500 animals. The first three variables were different linear dimensions, measured in centimetres, while the fourth variable was the weight of the animal measured in grammes. The sample covariance matrix was calculated and its four eigenvalues were found to be 14.1, 4.3, 1.2 and 0.4. The corresponding first and second eigenvectors were:

$$\mathbf{a}_1^T = [0.39, 0.42, 0.44, 0.69]$$
$$\mathbf{a}_2^T = [0.40, 0.39, 0.42, -0.72]$$

Report briefly on these results.

Suppose the data were stored by recording the eigenvalues and eigenvectors together with the 500 values of the first and second principal components and the mean values for the original variables. Show how to reconstruct the original covariance matrix and an approximation to the original data.

4.8 This exercise demonstrates that regression equations are scale-invariant, in contrast to principal components. A random sample size n, $[(x_i, y_i); i = 1, \ldots, n]$, is taken on two variables. Write down the least-squares estimators for α and β when fitting a linear regression line, $E(Y|x) = \alpha + \beta x$. If the x-variable is transformed to a new variable, denoted by $x^* = c + dx$, where c, d are constants, show that an equivalent regression line results.

CHAPTER FIVE

Factor analysis

5.1 Introduction

Factor analysis (abbreviated to FA) has somewhat similar aims to principal component analysis (PCA) in that it is a variable-directed technique which is appropriate when the variables arise 'on an equal footing'. The idea is to derive new variables called *factors* which will hopefully give us a better understanding of the data. But whereas PCA produces an orthogonal transformation of the variables which depends on no underlying model, FA is based on a proper statistical model and is more concerned with explaining the covariance structure of the variables than with explaining the variances. Any variance which is unexplained by the common factors can be described by residual 'error' terms. Despite several basic differences, it is an unfortunate fact that FA and PCA are often confused by non-statisticians.

The basic ideas of FA were suggested around the turn of the century by Francis Galton and Charles Spearman among others, and originated mainly from the efforts of psychologists to gain a better understanding of 'intelligence'. Intelligence tests customarily contain a large variety of questions which depend to a greater or lesser extent on verbal ability, mathematical ability, memory, etc. FA was developed to analyse these test scores so as to determine if 'intelligence' is made up of a single underlying general factor or of several more limited factors measuring attributes like 'mathematical ability'. The results are still debatable.

Most applications of FA have been in psychology and the social sciences. As one sociological example, suppose that information is collected from a wide range of people as to their occupation, type of education, whether or not they own their own home, and so on. Then one might ask if the concept of social class is multidimensional or if it is possible to construct a single 'index' of class from the data. In other words, we ask if their is a single underlying factor.

While FA overcomes some of the objections to PCA, it unfortunately introduces new problems, as discussed in Section 5.4. In particular, the FA model described below requires one to make a number of questionable assumptions about the data.

5.2 The factor-analysis model

Suppose we make observations on p variables, $X_1, X_2, \ldots, X_p$, which have mean vector $\boldsymbol{\mu}$ and covariance matrix Σ. As we are interested in explaining the covariance structure of the variables, we may without loss of generality assume that $\boldsymbol{\mu} = \mathbf{0}$. It will also be convenient to assume that Σ is of full rank p.

The FA model assumes that there are m underlying factors (where $m < p$) which we denote by $f_1, f_2, \ldots, f_m$, and that each observed variable is a linear function of these factors together with a residual variate, so that

$$X_j = \lambda_{ji} f_1 + \ldots + \lambda_{jm} f_m + e_j \qquad j = 1, \ldots, p \tag{5.1}$$

In Equation (5.1), the weights $\{\lambda_{jk}\}$ are usually called the *factor loadings*, so that λ_{jk} is the loading of the jth variable on the kth factor. The variate e_j describes the residual variation specific to the jth variable.

For obvious reasons the factors $\{f_i\}$ are often called the *common* factors, while the residual variates $\{e_j\}$ are often called the *specific* factors. The reader may compare Equation (5.1) with the inverse transformation (4.13) arising from PCA.

The factor analyst usually makes a number of assumptions about model (5.1). The specific factors are assumed to be independent of one another and of the common factors. It is also usually assumed that the common factors are independent of one another, though this assumption is sometimes relaxed when the factors are later rotated. As we have assumed the X's to have zero mean, it is also convenient to assume that the factors all have zero mean. Looking at Equation (5.1), we see that there is an arbitrary scale factor related to each common factor and so it is customary to choose the common factors so that each has unit variance. But the variances of the specific factors may vary, and we denote the variance of e_j by ψ_j.

It is also customary to assume that the common factors and the specific factors each have a multivariate normal distribution. This implies that $\mathbf{X}$ is also multivariate normal, where $\mathbf{X}^{\mathrm{T}} = [X_1, X_2, \ldots, X_p]$.

The large number of assumptions which have to be made in setting up the FA model is one of the drawbacks to the method. Indeed, Lawley and Maxwell (1971, p. 38) stress that the model is 'useful only as an approximation to reality' and 'should not be taken too seriously'.

Note that Equation (5.1) describes a relationship between random variables. With actual observations we write

$$\begin{aligned} x_{rj} &= r\text{th observation on variable } j \\ &= \sum_{k=1}^{m} \lambda_{jk} f_{rk} + e_{rj} \end{aligned}$$

where f_{rk} is the score of the kth common factor for the rth observation and e_{rj} is the value of the jth specific factor for the rth observation.

The model (5.1) is usually written in matrix notation in the form

$$\mathbf{X} = \Lambda \mathbf{f} + \mathbf{e} \tag{5.2}$$

where $\mathbf{f}^{\mathrm{T}} = [f_1, f_2, \ldots, f_m]$
$\mathbf{e}^{\mathrm{T}} = [e_1, e_2, \ldots, e_p]$

and

$$\Lambda = \begin{bmatrix} \lambda_{11} & \lambda_{12} & \cdots & \lambda_{1m} \\ \lambda_{21} & & & \lambda_{2m} \\ \vdots & & & \vdots \\ \lambda_{p1} & & & \lambda_{pm} \end{bmatrix}$$

Here Λ is of order $(p \times m)$ and should not be confused with the diagonal matrix of eigenvalues which was denoted by the same symbol in Chapter 4.

From Equation (5.1), using the independence of different factors, we have

$$\begin{aligned} \operatorname{Var}(X_j) &= \lambda_{j1}^2 + \lambda_{j2}^2 + \ldots + \lambda_{jm}^2 + \operatorname{Var}(e_j) \\ &= \sum_{k=1}^{m} \lambda_{jk}^2 + \psi_j \end{aligned} \tag{5.3}$$

since the common factors have unit variance. The part of the variance explained by the common factors, namely $\sum_{k=1}^{m} \lambda_{jk}^2$, is called the *communality* of the jth variable.

From Equation (5.1) we also find

$$\operatorname{Cov}(X_i, X_j) = \sum_{k=1}^{m} \lambda_{ik}\lambda_{jk}$$

Thus the covariance matrix of $\mathbf{X}$, which we denote as usual by Σ, is given by

$$\Sigma = \Lambda\Lambda^{\mathrm{T}} + \Psi \tag{5.4}$$

where

$$\Psi = \begin{bmatrix} \psi_1 & 0 & \cdots & 0 \\ 0 & \psi_2 & \cdots & 0 \\ \vdots & \vdots & & \\ 0 & 0 & \cdots & \psi_p \end{bmatrix}$$

Equation (5.4) is of crucial importance in FA. It demonstrates that the factors 'explain' the off-diagonal terms of Σ (namely the covariances) exactly, since Ψ is diagonal. It also establishes that finding the factor loadings is essentially equivalent to factorizing the covariance matrix of $\mathbf{X}$ in this particular way, with the added condition that the diagonal elements of Ψ must be non-negative.

Given a particular Σ, we now consider what conditions are necessary for the factorization in Equation (5.4) to exist, and, if it does, whether it is unique. Unfortunately, these are not easy questions to answer. The total number of parameters we have to estimate is the number of factor loadings, namely pm, plus the number of communal variances, namely p. Now there are $\frac{1}{2}p(p+1)$ separate variances and covariances in Σ, so that, equating corresponding elements of the matrices on both sides of Equation (5.4), we have $\frac{1}{2}p(p+1)$ equations. We generally require the number of parameters to be less than the number of equations, so that $(pm+p) < \frac{1}{2}p(p+1)$ or $m < \frac{1}{2}(p-1)$. In other words, m should be fairly small compared with p. But this does not guarantee a solution will exist.

Consider the case $m = 1$. Then Λ is a $(p \times 1)$ column vector, where $\Lambda^{\mathrm{T}} = [\lambda_{11}, \lambda_{21}, \ldots, \lambda_{p1}]$. Then Equation (5.4) implies that the off-diagonal terms of Σ must be of the form

$$\begin{bmatrix} - & \lambda_{11}\lambda_{21} & \lambda_{11}\lambda_{31} & \ldots & \lambda_{11}\lambda_{p1} \\ \lambda_{21}\lambda_{11} & - & \lambda_{21}\lambda_{31} & \ldots & \lambda_{21}\lambda_{p1} \\ \vdots & & & & \\ \lambda_{p1}\lambda_{11} & \ldots & & & - \end{bmatrix}$$

If we choose to analyse the correlation matrix rather than the covariance matrix, then the above result implies that

$$\frac{\text{Correlation}\,(X_i, X_k)}{\text{Correlation}\,(X_j, X_k)} = \frac{\lambda_{i1}}{\lambda_{j1}}$$

so that the off-diagonal terms in the correlation matrix must be such that the corresponding elements in any two rows (or in any two columns) are in the same ratio. In other words if the top row of the correlation matrix is $[1, \rho_{12}, \rho_{13}, \ldots, \rho_{1p}]$, then for example, the second row must be of the form $[\rho_{12}, 1, k\rho_{13}, \ldots, k\rho_{1p}]$, where $k = \lambda_{21}/\lambda_{11}$. This pattern should also hold approximately in the sample correlation matrix, and it was this feature in a correlation matrix of examination scores that was noticed by Charles Spearman about 1900 and led him to propose a one-factor model.

It is easy to see that the off-diagonal elements of Σ will not always have the above pattern, so that an acceptable one-factor solution will not always exist. Lawley and Maxwell (1971, Chapter 2) consider two examples of correlation matrices which lead to negative estimates of communal variances or imaginary solutions for the $\{\lambda_{j1}\}$ when one tries to find a one-factor solution.

If a solution does exist in the case $m = 1$, then it will usually, though not always, be unique. But if $m > 1$ and a solution exists, then it is easy to show that the solution is not unique. Using Equation (5.4), we see that any orthogonal rotation of the factors in the relevant m-space will give us a new set of factors which will also satisfy Equation (5.4). Let T be an orthogonal matrix of order $(m \times m)$. Then

$$\begin{aligned}(\Lambda T)(\Lambda T)^{\mathrm{T}} &= \Lambda T T^{\mathrm{T}} \Lambda^{\mathrm{T}} \\ &= \Lambda \Lambda^{\mathrm{T}}\end{aligned} \tag{5.5}$$

so that although the loadings in Λ and (ΛT) are different, their ability to generate the given covariances of **X** is the same.

In order to obtain a unique solution in the case $m > 1$, the methods used to extract factors from sample data effectively put extra constraints on Λ as discussed below. But the lack of uniqueness implied by Equation (5.5) means that one can always rotate a solution to an alternative solution. The possibility of rotating solutions so as to assist in their interpretation is attractive at first sight, but is also potentially dangerous (see Section 5.4).

5.3 Estimating the factor loadings

The parameters of the FA model, including the factor loadings and the error variances, are nearly always unknown and need to be estimated from the sample data. The sample covariance matrix is occasionally used, but it is much more common to work with the sample correlation matrix. Then the X-variables in Equations (5.1) and (5.2) refer to scaled variables having zero mean and unit variance. It then follows from Equation (5.3) that

$$1 = \sum_{k=1}^{m} \lambda_{jk}^2 + \psi_j$$

In the early days of factor analysis, a variety of iterative methods were used to estimate the factor loadings. These involved subjective judgement, such as guessing the communalities, with the result that different researchers could analyse the same data and find entirely different factors. (This still happens today!) One popular method was the principal-factor method. This chooses the first factor so as to account for as much as possible of the communal variance, the second factor to account for as much as possible of the remaining communal variance, and so on. The method requires suitable estimates of the communalities. If they are chosen to be unity, then the method reduces to principal component analysis. In particular, if we choose to fit a one-factor model by minimizing the total residual sum of squares, then the first factor turns out to be the first principal component of the data.

Then in 1940, a major step forward was made by D. N. Lawley, who developed the maximum-likelihood equations. These are fairly complicated and difficult to solve, but recent computational advances, particularly by K. G. Jöreskog, have made maximum-likelihood estimation a practical proposition, and computer programs are now becoming widely available. However, some programs labelled as FA programs can still be found which essentially carry out a PCA.

The constraint on Λ which is used in the maximum-likelihood approach is to require that $\Lambda^T \Psi^{-1} \Lambda$ be diagonal with elements arranged in descending order of magnitude. This constraint is related to the condition imposed in the principal-factor method. One useful advantage of maximum-likelihood estimates is that they turn out to be scale-invariant, and so do not have the scaling problems associated with principal component analysis, as discussed in Section 4.4. If we make a linear transformation of one of the observed variates, then the corresponding factor loadings simply change by a multiple depending on the standard deviations of the original and scaled variables (see Equation (5.1)). In particular the loadings extracted from the sample covariance matrix of a set of data, when multiplied by the reciprocal of the appropriate standard deviation, give the loadings for the sample correlation matrix.

As already remarked, there is no guarantee that an acceptable solution exists and there appear to be no general conditions on the form of the sample covariance matrix which will ensure that it can be factorized as in Equation (5.4). In addition, if one tries to find maximum-likelihood estimates, there is no guarantee that the iterative procedure which is involved will converge to an acceptable solution. In particular, one or more of the estimates of the residual variances, ψ_j, may tend to zero and become negative if allowed to do so. The residual variances are usually constrained so that $\psi_j \geq \varepsilon$ for all j, where ε is a 'small' positive value. A solution which produces one or more of the residual variances to be equal to ε is called an *improper* solution. This may arise because of sampling errors or because the FA model is inappropriate.

When a set of factors has been derived, they are not always easy to interpret. Various methods have been proposed for *rotating* the factors to find new ones which may be easier to interpret. The usual aim of rotation methods, such as the *varimax* method, is to make the loadings 'large' or 'small' so that most variables have a high loading on a small number of factors. Most rotation methods involve an orthogonal rotation as in Equation (5.5), but some rotation methods, like the *promax* method, allow the factors to become correlated rather than independent, although this does not necessarily make interpretation easier. Kendall (1975, Chapter 4) prefers to carry out a cluster analysis of the variables rather than allow correlated factors.

We shall not attempt to derive the maximum-likelihood equations, discuss numerical procedures for solving them, or describe rotation methods. A full treatment of estimation problems in FA is given in the books by Lawley and Maxwell (1971), Harman (1976) and Jöreskog *et al.* (1976) and there seems little point in duplicating this extensive work here. This is particularly true in view of doubts about the general usefulness of factor analysis as discussed in the next section.

Another topic we shall not discuss here is that of *confirmatory factor analysis* (e.g., Maxwell, 1977, Chapter 6) which is used for testing a hypothesis that there is a given pattern of factor loadings.

5.4 Discussion

The basic distinction between FA and PCA is not always made clear in the literature. We have seen that, whereas PCA is a technique for finding an orthogonal rotation of the variables, FA depends on a 'proper' statistical model. The two methods are similar in one respect, namely that they are both pointless if all the observed variables are approximately uncorrelated; FA because it has nothing to explain, and PCA because it would simply reveal components which are similar to the original variables.

Suppose we analyse the same set of data by means of PCA and FA. The question arises as to whether the factors in an m-factor model will have loadings similar to the component correlations of the first m principal components. (The component correlations are the weights which arise when one chooses the components to have unit variance – see Section 4.3.7.) The answer is that there will sometimes be good agreement but usually there will not. Unfortunately, it is hard to give general rules as to when there will be reasonable agreement. It depends to some extent on what rotations, if any, are carried out in the factor analysis, and whether m is chosen to be the number of eigenvalues which lie above a clear breakpoint in the sequence of eigenvalues of S (or R). Some people prefer to use PCA even though no 'error' structure is implied. But, if the observations are subject to random perturbations, then it should be remembered that the 'errors' will inflate the variances of the first few components as well as those of the remaining components.

In view of the tremendous current interest in FA, particularly in the sociological sciences, the reader may be surprised that our treatment of the subject is so brief. The reason for this is simple. Although we regard PCA as moderately useful (though not as useful as some people think), we find ourselves in sympathy with the growing group of statisticians who doubt if FA is worth using except in a few particular types of application. For example, Hills (1977) has said that FA is not 'worth the time necessary to understand it and carry it out'. He goes on to say that he regards FA as an 'elaborate way of doing something which can only ever be crude, namely picking out clusters of inter-related variables, and then finding some sort of average of the variables in a cluster in spite of the fact that the variables may be measured on different scales.'

Admittedly FA does have some advantages over PCA, particularly in that maximum-likelihood estimation overcomes the scaling problem and that a proper statistical model, with an error structure, is involved. But there are many drawbacks to FA:

(*a*) A large number of assumptions have to be made in setting up the FA model. These assumptions are not always realistic in practice.

(*b*) An even more basic assumption in the FA model is that the factors exist at all; The concept of a set of underlying *unobservable* variables is one which may be reasonable in some situations, particularly in psychological research, but is not appealing in many other practical situations.

(*c*) The FA model also assumes knowledge of m, the number of factors. In practice, m is often unknown and different values may be tried sequentially, starting with $m = 1$. But it is not easy to select the 'correct' value of m. Although a test is available (see Lawley and Maxwell, 1971, Section 4.4), it is rather complicated and depends on the model assumptions, so that external considerations are often used to select the value of m. It is somewhat disturbing to find that the form of the factors (as determined by which loadings are 'large') may change completely as m changes. In contrast, the components derived in PCA are unique (except where there are equal eigenvalues) and so stay the same as one varies the number of components which are thought to be worth including.

(*d*) Even for a given value of *m*, the factors are not unique as different methods of rotation may produce sets of factors which look quite different. Although rotation is mathematically respectable, the danger here is that the analyst may go on trying different values of *m* and different methods of rotation until he gets the answer he is looking for. The lack of uniqueness introduced by allowing rotation also has the drawback that different investigators may use different rotations on the same set of data and get apparently different results. In addition, it is unusual to get repeatable results on replicate samples.

(*e*) In PCA it is easy to calculate component scores for an individual using Equation (4.12), and these can readily be used in subsequent analyses. But the FA model in Equation (5.2) has no obvious inverse and it is not so easy to estimate factor scores from observed data, though two methods of estimating factor scores are described by Lawley and Maxwell (1971, Chapter 8). This makes it more difficult to use FA than PCA in follow-up analyses.

(*f*) We have already seen that principal components are often difficult to interpret. The same is often true of estimated factors. Now a complicated technique like factor analysis cannot be regarded as useful unless it leads to a deeper understanding of the data. We know of few cases where this has happened apart from psychology, where certain underlying factors have been reported in the literature and verified in later studies. In those situations where factors do have an obvious interpretation, the relationships are often obvious from a visual inspection of the correlation matrix as discussed in Sections 3.2 and 4.5.1.

For a further discussion on the controversy surrounding factor analysis, we recommend Chapter 13 of Blackith and Reyment (1971). These authors consider a number of scientific examples and conclude that PCA is preferable to FA. Even in psychology, the latter method is not without opponents. Indeed Blackith and Reyment wonder (p. 201) if the method has 'persisted precisely because it allows the experimenter to impose his preconceived ideas on the raw data'.

In view of the disadvantages listed above, we recommend that FA should not be used in most practical situations, although we recognize that this view will be controversial.

PART THREE

Procedures based on the multivariate normal distribution

This part of the book is concerned with multivariate procedures which are extensions of the traditional univariate procedures usually developed from normal distribution sampling theory. The development here is also traditional. Chapter 6 cover the required multivariate normal distribution theory. The following three chapters present the procedures. Chapter 7 is concerned with one- and two-sample procedures, including discriminant analysis. Chapter 8 deals with the multivariate analysis of variance and Chapter 9 with the multivariate analysis of covariance.

However, Chapters 7–9 present the procedures not from a theoretical point of view but rather from the reader's assumed background familiarity with the analogous univariate procedures. Reference back to Chapter 6 is made only to justify distributional assertions. If the reader is prepared to accept these assertions, then Chapters 7–9 may be taken on their own as a practical course in methodology.

Furthermore, it can be argued that the methods presented have intrinsic appeal as data-analytic methods and, in this role, produce valuable and sensible summary and descriptive statistics. By this argument the methodology can be used without requiring any distributional or sampling assumptions.

CHAPTER SIX
The multivariate normal distribution

6.1 Introduction

The multivariate normal distribution was briefly introduced in Chapter 2. In this chapter we consider its properties in some detail since the estimation of its parameters is the source of many standard multivariate statistical methods. In addition, we shall meet a number of derived distributions of fundamental importance.

The mathematical content is necessarily rather more difficult than that in the rest of the book and the reader may prefer to accept some of the results at a first reading without going into the derivations in detail.

6.2 Definition of the multivariate normal distribution

Suppose $\mathbf{X}$ is a p-dimensional random variable with $E(\mathbf{X}) = \boldsymbol{\mu}$ and $\text{Var}(\mathbf{X}) = \Sigma$. Then if $\mathbf{a}$ is a $(p \times 1)$ vector of constants, $U = \mathbf{a}^T\mathbf{X}$ is a scalar random variable. It was shown in Chapter 2 that $E(U) = \mathbf{a}^T\boldsymbol{\mu}$ and $\text{Var}(U) = \mathbf{a}^T\Sigma\mathbf{a}$. We shall call U a *linear compound* of $\mathbf{X}$. In fact, the relation $U = \mathbf{a}^T\mathbf{X}$ defines a linear functional which maps the vector values taken by $\mathbf{X}$ on to the scalar values taken by U.

An interesting result due to Cramer and Wold is that the distribution of a p-dimensional random variable is completely determined by the (univariate) distributions of all its linear compounds.

The result follows by considering $\phi(z)$, the characteristic function of a scalar random variable U. By definition,

$$\phi(z) = E[\exp\{izU\}]$$

If $U = \mathbf{t}^T\mathbf{X}$, a linear compound of $\mathbf{X}$, we may write

$$\phi(z, \mathbf{t}) = E[\exp\{iz\mathbf{t}^T\mathbf{X}\}]$$

Setting $z = 1$, we have

$$\phi(\mathbf{t}) = E[\exp\{i\mathbf{t}^T\mathbf{X}\}]$$

This is a function of the p elements of $\mathbf{t}$ and as such is, by definition, the characteristic function of the multivariate distribution of $\mathbf{X}$. Hence, since a distribution is uniquely defined by its characteristic function, we may define the distribution of a p-dimensional random variable by specifying the distributions of all its linear compounds. This approach will be used to give a more general definition of the multivariate normal distribution than that provided by Equation (2.13).

Definition. A p-dimensional random variable $\mathbf{X}$ *is said to have a multivariate normal distribution if and only if every linear compound of* $\mathbf{X}$ *has a univariate normal distribution.* □

Now each component variate of $\mathbf{X}$ is a linear compound of $\mathbf{X}$ (for instance $X_1 = [1, 0, 0, \ldots, 0]\mathbf{X}$) and, by the definition, has a univariate normal distribution. Hence, we may deduce that $E(\mathbf{X}) = \boldsymbol{\mu}$ and $\mathrm{Var}(\mathbf{X}) = \Sigma$ exist since their elements are the means, variances and covariances of one-dimensional (scalar) normal variates and these are known to exist.

It was shown in Section 2.2 that Σ is positive semidefinite and the above definition imposes no further constraint on Σ. However, the density function given in Equation (2.13) requires Σ to be of full rank (hence positive definite) to ensure the existence of Σ^{-1}. A multivariate normal distribution for which Σ^{-1} does not exist is called a *singular* or *degenerate normal distribution* and does not possess a density function. On the other hand, the characteristic function of the distribution always exists and may be derived by the approach used earlier in the section. We proceed as follows:

Suppose $U \sim \mathrm{N}(\mu, \sigma^2)$. The characteristic function of the distribution of U is given by

$$\phi(z) = \exp\{iz\mu - \tfrac{1}{2}z^2\sigma^2\}$$

Hence, if $U = \mathbf{t}^{\mathrm{T}}\mathbf{X}$, $U \sim \mathrm{N}(\mathbf{t}^{\mathrm{T}}\boldsymbol{\mu}, \mathbf{t}^{\mathrm{T}}\Sigma\,\mathbf{t})$

and
$$\phi(z, \mathbf{t}) = \exp\{iz\mathbf{t}^{\mathrm{T}}\boldsymbol{\mu} - \tfrac{1}{2}z^2\mathbf{t}^{\mathrm{T}}\Sigma\,\mathbf{t}\}$$

Setting $z = 1$, we find

$$\phi(\mathbf{t}) = \exp\{i\mathbf{t}^{\mathrm{T}}\boldsymbol{\mu} - \tfrac{1}{2}\mathbf{t}^{\mathrm{T}}\Sigma\,\mathbf{t}\} \qquad (6.1)$$

This is the characteristic function of the multivariate normal distribution. Note that the distribution is completely specified by $\boldsymbol{\mu}$ and Σ, justifying the use of the notation $\mathbf{X} \sim \mathrm{N}_p(\boldsymbol{\mu}, \Sigma)$.

An immediate consequence of the definition is that multivariate normality is preserved under a linear transformation. If $\mathbf{Y} = A^{\mathrm{T}}\mathbf{X}$, where A is a matrix of constants of order $(p \times m)$, then

$$\mathbf{Y} \sim \mathrm{N}_m(A^{\mathrm{T}}\boldsymbol{\mu}, A^{\mathrm{T}}\Sigma\,A) \qquad (6.2)$$

since any linear compound of $\mathbf{Y}$ is a linear compound of $\mathbf{X}$ and hence is normally distributed. The mean and variance follow from Equation (2.9).

We now turn to the interesting problem concerned with finding the transformation between the random vector $\mathbf{X} \sim \mathrm{N}_p(\boldsymbol{\mu}, \Sigma)$ and a random vector $\mathbf{U}$ whose components have independent standard normal distributions so that $\mathbf{U} \sim N_m(\mathbf{0}, I)$ where $m = \text{rank}(\Sigma)$ and I is necessarily of order m being the covariance matrix of a vector random variable of dimension m. This is the multivariate equivalent of transforming an arbitrary univariate normal distribution to the standard normal distribution and we shall find it useful in the subsequent theory.

Suppose, firstly, that Σ is of full rank. Then there exists a non-singular $(p \times p)$ matrix B such that $\Sigma = BB^T$. Now consider the transformation $(\mathbf{X} - \boldsymbol{\mu}) = B\mathbf{U}$.
If $\mathbf{U} \sim N_p(\mathbf{0}, I)$, $(\mathbf{X} - \boldsymbol{\mu}) \sim N_p(\mathbf{0}, BB^T)$, by Equation (6.2).
Hence $\mathbf{X} \sim N_p(\boldsymbol{\mu}, \Sigma)$
Conversely, since B^{-1} exists, the inverse transformation is represented by $\mathbf{U} = B^{-1}(\mathbf{X} - \boldsymbol{\mu})$.
Then, if $\mathbf{X} \sim N_p(\boldsymbol{\mu}, \Sigma)$

$$\begin{aligned} E(\mathbf{U}) &= \mathbf{0} \\ \text{Var}(\mathbf{U}) &= B^{-1}\Sigma(B^{-1})^T \qquad \text{by Equation (6.2)} \\ &= B^{-1}(BB^T)(B^T)^{-1} \\ &= I_p \end{aligned}$$

so that $\mathbf{U} \sim N_p(\mathbf{0}, I)$.

There are many such transformations since, for given Σ, B is not unique.

Let us return to the more general case where Σ is not necessarily of full rank. The argument to be used below depends on the following result in matrix algebra.

Suppose Σ is a $(p \times p)$ symmetric, positive semidefinite, real matrix of rank $m < p$. Then there exists a non-singular square matrix P such that

$$P\Sigma P^T = \begin{bmatrix} I_m & 0 \\ 0 & 0 \end{bmatrix}$$

Write

$$P = \begin{bmatrix} P_1 \\ P_2 \end{bmatrix}$$

where P_1 is $(m \times p)$ and of rank necessarily equal to m. It follows that $P_1\Sigma P_1^T = I_m$.
Also if $Q = P^{-1}$ and is partitioned so that $Q = [Q_1 \vdots Q_2]$, where Q_1 is $(p \times m)$ of rank m,

$$\Sigma = Q\begin{bmatrix} I_m & 0 \\ 0 & 0 \end{bmatrix}Q^T = Q_1Q_1^T$$

By the above result, given Σ of rank m, matrices P_1 and Q_1 exist such that $\Sigma = Q_1Q_1^T$ and $P_1\Sigma P_1^T = I_m$.

Then given $\mathbf{U} \sim \mathrm{N}_m(\mathbf{0}, I)$, the transformation $(\mathbf{X} - \boldsymbol{\mu}) = Q_1\mathbf{U}$ defines the p component variates of $(\mathbf{X} - \boldsymbol{\mu})$ as linear compounds of $\mathbf{U}$, so that $\mathbf{X} \sim \mathrm{N}_p(\boldsymbol{\mu}, Q_1Q_1^{\mathrm{T}})$, and we say that $\mathbf{X} \sim \mathrm{N}_p(\boldsymbol{\mu}, \Sigma)$ of rank m.

Conversely, given $\mathbf{X} \sim \mathrm{N}_p(\boldsymbol{\mu}, \Sigma)$, the transformation $\mathbf{U} = P_1(\mathbf{X} - \boldsymbol{\mu})$ defines the m component variates of $\mathbf{U}$ as linear compounds of $(\mathbf{X} - \boldsymbol{\mu})$ so that $\mathbf{U} \sim \mathrm{N}_m(\mathbf{0}, P_1\Sigma P_1^{\mathrm{T}})$, i.e., $\mathbf{U} \sim \mathrm{N}_m(\mathbf{0}, I)$.

One practical method which may be used to find suitable transformations uses the canonical reduction of Σ referred to in Section 1.6 on matrix algebra, specifically Equations (1.15) and (1.16). If Λ denotes the diagonal matrix of non-zero eigenvalues and C is the $(p \times m)$ matrix of rank m whose columns are the corresponding eigenvectors, $P_1 = \Lambda^{-1/2}C^{\mathrm{T}}$ and $Q_1 = C\Lambda^{1/2}$ satisfy the requirements of the theory. When Σ is of full rank, C is the $(p \times p)$ matrix of eigenvectors, $P_1 = P = \Lambda^{-1/2}C^{\mathrm{T}}$ and $Q_1 = Q = P^{-1} = B$. The method is then a principal component analysis modified so that the transformed variates have unit variance (see Section 4.3.7).

The results of the above discussion are formally presented in the following theorem:

Theorem 6.1. $\mathbf{X} \sim \mathrm{N}_p(\boldsymbol{\mu}, \Sigma)$ *of rank* $m \le p$ *if and only if* $\mathbf{X} = \boldsymbol{\mu} + B\mathbf{U}$, *where* $\mathbf{U} \sim \mathrm{N}_m(\mathbf{0}, I)$, $BB^{\mathrm{T}} = \Sigma$ *and* B *is a* $(p \times m)$ *matrix of rank* m. □

Corollary 6.1. *If* Σ *is of full rank,* B *is a* $(p \times p)$ *matrix of full rank and we may write* $\mathbf{U} = B^{-1}(\mathbf{X} - \boldsymbol{\mu})$. □

A slightly different proof is given in Rao (1973, Section 8a2(f)) and an alternative approach is used in his Section 9g1 leading to a method of finding suitable transformations based on Gram–Schmidt orthogonalization. The presentation in Anderson (1958) is essentially the same as that given above.

From the discussion above, we see that when $\mathbf{X}$ has a singular distribution of rank $m < p$, the component variates of $\mathbf{X}$ may be expressed as linear combinations of m independent standard normal variates and that one way of finding a suitable transformation is based on the eigenvectors and eigenvalues of Σ. The $(p - m)$ eigenvalues which are zero imply that there exist $(p - m)$ linear relationships among the component variates of $\mathbf{X}$. The eigenvectors defining the relationships are not unique (unless $p - m = 1$) and we may present them in many different ways (Sections 4.3.3 and 4.3.5).

For a simple illustrative example of a degenerate distribution, consider the three-dimensional random variable $\mathbf{X}$ where the component variates X_1, X_2 and X_3 are, respectively, the lengths, breadths and perimeters of random rectangles. Note that the linear relationship $2X_1 + 2X_2 - X_3 = 0$ applies to each individual member of the population of rectangles. This is an example of a *structural relation*. Although we choose to record a three-dimensional vector observation, the variation is essentially two-dimensional, and $\mathrm{rank}(\Sigma) = 2$.

When Σ is of full rank, the probability density function exists and can easily be derived using Corollary 6.1 as follows:

Suppose $\mathbf{X} \sim \mathrm{N}_p(\boldsymbol{\mu}, \Sigma)$ of rank p. Then we can find a transformation

$$\mathbf{U} = B^{-1}(\mathbf{X} - \boldsymbol{\mu})$$

such that $BB^{\mathrm{T}} = \Sigma$ and $\mathbf{U} \sim \mathrm{N}_p(\mathbf{0}, I)$.

Since the component variates of $\mathbf{U}$ are independent standard normal variates, the density function of $\mathbf{U}$ is given by

$$\begin{aligned} g(\mathbf{u}) &= \prod_{j=1}^{p} (2\pi)^{-1/2} \exp\{-\tfrac{1}{2}u_j^2\} \\ &= (2\pi)^{-p/2} \exp\{-\tfrac{1}{2}\mathbf{u}^{\mathrm{T}}\mathbf{u}\} \end{aligned}$$

and the density function of $\mathbf{X}$ is then

$$f(\mathbf{x}) = g(\mathbf{u}(\mathbf{x}))|J|$$

where $|J|$ is the Jacobian of the transformation. For the linear transformation

$$\mathbf{U} = B^{-1}(\mathbf{X} - \boldsymbol{\mu})$$

$$|J| = |B^{-1}| = |B|^{-1} = (|B|\,|B^{\mathrm{T}}|)^{-1/2} = |\Sigma|^{-1/2}$$

and $\mathbf{u}^{\mathrm{T}}\mathbf{u} = (\mathbf{x} - \boldsymbol{\mu})^{\mathrm{T}}(B^{-1})^{\mathrm{T}}B^{-1}(\mathbf{x} - \boldsymbol{\mu})$

$$\begin{aligned} &= (\mathbf{x} - \boldsymbol{\mu})^{\mathrm{T}}(BB^{\mathrm{T}})^{-1}(\mathbf{x} - \boldsymbol{\mu}) \\ &= (\mathbf{x} - \boldsymbol{\mu})^{\mathrm{T}}\Sigma^{-1}(\mathbf{x} - \boldsymbol{\mu}) \end{aligned}$$

Hence

$$f(\mathbf{x}) = (2\pi)^{-p/2}|\Sigma|^{-1/2}\exp\{-\tfrac{1}{2}(\mathbf{x} - \boldsymbol{\mu})^{\mathrm{T}}\Sigma^{-1}(\mathbf{x} - \boldsymbol{\mu})\} \qquad (6.3)$$

which is the form given in Equation (2.13).

As we have seen, the density does not exist when $\mathrm{rank}(\Sigma) = m < p$ because there exist structural relations among the component variates of $\mathbf{X}$. However we can use Theorem 6.1 to transform to an underlying basic m-dimensional random variable whose density does exist.

Note that the definition of the multivariate normal distribution given in this chapter implies the definition given in Chapter 2, but the converse does not hold when Σ is not of full rank.

6.3 Properties of the multivariate normal distribution

In this section we shall establish a number of additional useful properties of the normal distribution.

(i) If $\mathbf{X} \sim \mathrm{N}_p(\boldsymbol{\mu}, \Sigma)$ of rank p, so that Σ^{-1} exists, then

$$(\mathbf{X} - \boldsymbol{\mu})^{\mathrm{T}}\Sigma^{-1}(\mathbf{X} - \boldsymbol{\mu}) \sim \chi^2(p) \qquad (6.4)$$

This is a generalization of a property of the univariate normal distribution for, when $p = 1$, Equation (6.4) reduces to the well-known result that $[(X - \mu)/\sigma]^2 \sim \chi^2(1)$. By Corollary 6.1 we may write $\mathbf{U} = B^{-1}(\mathbf{X} - \boldsymbol{\mu})$, $BB^{\mathrm{T}} = \Sigma$, such that $\mathbf{U} \sim \mathrm{N}_p(\mathbf{0}, I)$. Then $(\mathbf{X} - \boldsymbol{\mu})^{\mathrm{T}}\Sigma^{-1}(\mathbf{X} - \boldsymbol{\mu}) = \mathbf{U}^{\mathrm{T}}\mathbf{U} = \sum_{j=1}^{p} U_j^2$, where the U_j are independently distributed standard normal variates. Hence the result.

By writing $\mathbf{U} = B^{-1}(\mathbf{X} - \boldsymbol{\mu}_0)$, it follows by a similar argument that

$$(\mathbf{X} - \boldsymbol{\mu}_0)^{\mathrm{T}}\Sigma^{-1}(\mathbf{X} - \boldsymbol{\mu}_0) \sim \chi^2(p; \delta^2)\dagger \tag{6.5}$$

where $\delta^2 = (\boldsymbol{\mu} - \boldsymbol{\mu}_0)^{\mathrm{T}}\Sigma^{-1}(\boldsymbol{\mu} - \boldsymbol{\mu}_0)$. Here the U_j are independently distributed normal variates with unit variance but with non-zero means, so that $\sum_j U_j^2$ has a non-central chi-square distribution. Note that to obtain the non-centrality parameter, $\mathbf{X}$ in the quadratic form in Equation (6.5) is replaced by its mean $\boldsymbol{\mu}$.

The next three results are concerned with the distributions of subsets of the p component variates of $\mathbf{X}$. In particular, it will be shown that the marginal and conditional distributions of a multivariate normal are also normal.

Suppose, after any necessary re-ordering, $\mathbf{X}$ is partitioned as

$$\mathbf{X} = \begin{bmatrix} \mathbf{X}_1 \\ \mathbf{X}_2 \end{bmatrix} \qquad \text{where } \mathbf{X}_1 \text{ is } (q \times 1),\ q < p$$

with corresponding partitions

$$\boldsymbol{\mu} = \begin{bmatrix} \boldsymbol{\mu}_1 \\ \boldsymbol{\mu}_2 \end{bmatrix}, \quad \Sigma = \begin{bmatrix} \Sigma_{11} & \Sigma_{12} \\ \Sigma_{21} & \Sigma_{22} \end{bmatrix}$$

Note that Σ_{11} and Σ_{22} are symmetric positive semidefinite matrices of order q and $p - q$ respectively, and that $\Sigma_{12} = \Sigma_{21}^{\mathrm{T}}$ is $q \times (p - q)$.

(ii) The *marginal distribution* of $\mathbf{X}_1$ is $\mathrm{N}_q(\boldsymbol{\mu}_1, \Sigma_{11})$.

The multivariate normality of $\mathbf{X}_1$ is established by noting that linear compounds of $\mathbf{X}_1$ are also linear compounds of $\mathbf{X}$ and so have univariate normal distributions. Note that the special case of this result when $q = 1$ has already been established when it was noted earlier that each component variate of $\mathbf{X}$ has a univariate normal distribution.

(iii) $\mathbf{X}_1$ and $\mathbf{X}_2$ are independently distributed if and only if $\Sigma_{12} = 0$.

To establish this result, note that $\Sigma_{12} = 0$ is a necessary and sufficient condition that, for any $\mathbf{t}$,

$$\mathbf{t}^{\mathrm{T}}\Sigma\,\mathbf{t} = \mathbf{t}_1^{\mathrm{T}}\Sigma_{11}\mathbf{t}_1 + \mathbf{t}_2^{\mathrm{T}}\Sigma_{22}\mathbf{t}_2, \qquad \text{where } \mathbf{t}^{\mathrm{T}} = [\mathbf{t}_1^{\mathrm{T}}, \mathbf{t}_2^{\mathrm{T}}]$$

and hence (see Equation (6.1)) that

$$\phi(\mathbf{t}) = \phi_1(\mathbf{t}_1)\phi_2(\mathbf{t}_2)$$

where $\phi(\mathbf{t})$, $\phi_1(\mathbf{t}_1)$ and $\phi_2(\mathbf{t}_2)$ are the characteristic functions of the distributions of $\mathbf{X}$, $\mathbf{X}_1$ and $\mathbf{X}_2$ respectively. The result follows since such factorization of the characteristic function is a necessary and sufficient condition for independence of distributions.

(iv) If Σ_{22} is of full rank so that Σ_{22}^{-1} exists, the *conditional distribution* of $\mathbf{X}_1$ given $\mathbf{X}_2 = \mathbf{x}_2$ is multivariate normal with

$$E(\mathbf{X}_1|\mathbf{X}_2 = \mathbf{x}_2) = \boldsymbol{\mu}_1 + \Sigma_{12}\Sigma_{22}^{-1}(\mathbf{x}_2 - \boldsymbol{\mu}_2)$$
$$\mathrm{Var}(\mathbf{X}_1|\mathbf{X}_2 = \mathbf{x}_2) = \Sigma_{11} - \Sigma_{12}\Sigma_{22}^{-1}\Sigma_{21}$$

† $\chi^2(p; \delta^2)$ denotes a non-central χ^2 variate with non-centrality parameter δ^2 such that $E[\chi^2(p; \delta^2)] = p + \delta^2$. The corresponding central χ^2 variate, when $\delta^2 = 0$, is denoted by $\chi^2(p)$.

Consider the q-dimensional random variable $\mathbf{Y} = \mathbf{X}_1 + M\mathbf{X}_2$, where M is a matrix of constants (of order $q \times (p-q)$) to be determined. Then $\mathbf{Y}$ is normally distributed since linear compounds of $\mathbf{Y}$ are linear compounds of $\mathbf{X}$ and so have univariate normal distributions.
Now $\text{Cov}(\mathbf{Y}, \mathbf{X}_2) = \text{Cov}((\mathbf{X}_1 + M\mathbf{X}_2), \mathbf{X}_2) = \Sigma_{12} + M\Sigma_{22}$, so that $\text{Cov}(\mathbf{Y}, \mathbf{X}_2) = 0$ if $M = -\Sigma_{12}\Sigma_{22}^{-1}$. It follows from (iii) that $\mathbf{Y}$ and $\mathbf{X}_2$ will then have independent normal distributions. Consequently, the conditional distribution of $\mathbf{Y}|\mathbf{X}_2 = \mathbf{x}_2$ will be the same as the marginal distribution of $\mathbf{Y}$.
For the marginal distribution of $\mathbf{Y}$, we have

$$\mathbf{Y} = \mathbf{X}_1 - \Sigma_{12}\Sigma_{22}^{-1}\mathbf{X}_2 = [I_q \,\vdots\, -\Sigma_{12}\Sigma_{22}^{-1}]\mathbf{X}$$
$$E(\mathbf{Y}) = \boldsymbol{\mu}_1 - \Sigma_{12}\Sigma_{22}^{-1}\boldsymbol{\mu}_2$$
$$\text{Var}(\mathbf{Y}) = [I_q \,\vdots\, -\Sigma_{12}\Sigma_{22}^{-1}]\Sigma\begin{bmatrix} I_q \\ -\Sigma_{22}^{-1}\Sigma_{21} \end{bmatrix}$$
$$= \Sigma_{11} - \Sigma_{12}\Sigma_{22}^{-1}\Sigma_{21}$$

For the conditional distribution of $\mathbf{Y}|\mathbf{X}_2 = \mathbf{x}_2$ ($\mathbf{x}_2$ is a constant),

$$(\mathbf{Y}|\mathbf{X}_2 = \mathbf{x}_2) = \mathbf{X}_1 - \Sigma_{12}\Sigma_{22}^{-1}\mathbf{x}_2$$
$$E(\mathbf{Y}|\mathbf{X}_2 = \mathbf{x}_2) = E(\mathbf{X}_1|\mathbf{X}_2 = \mathbf{x}_2) - \Sigma_{12}\Sigma_{22}^{-1}\mathbf{x}_2$$
$$\text{Var}(\mathbf{Y}|\mathbf{X}_2 = \mathbf{x}_2) = \text{Var}(\mathbf{X}_1|\mathbf{X}_2 = \mathbf{x}_2)$$

The result follows by equating the means and variances of the two distributions.

For the special case when $q = 1, \mathbf{X}_1 = X_1$, the first component variate of $\mathbf{X}$, and $\Sigma_{11} = \sigma_{11}$. It follows that

$$E(X_1|\mathbf{X}_2 = \mathbf{x}_2) = \mu_1 + \Sigma_{12}\Sigma_{22}^{-1}(\mathbf{x}_2 - \boldsymbol{\mu}_2) \tag{6.6}$$

which is of the form

$$E(X_1|\mathbf{X}_2 = \mathbf{x}_2) = \mu_1 + \beta_2(x_2 - \mu_2) + \ldots + \beta_p(x_p - \mu_p) \tag{6.7}$$

and is the *regression function* of X_1 on $X_2, X_3 \ldots X_p$.
The *variance about the regression function* may be written

$$\text{Var}(X_1|\mathbf{X}_2 = \mathbf{x}_2) = \sigma_{11} - \Sigma_{12}\Sigma_{22}^{-1}\Sigma_{21} \qquad \text{a } \textit{scalar}$$
$$= |\sigma_{11} - \Sigma_{12}\Sigma_{22}^{-1}\Sigma_{21}|$$
$$= \frac{|\Sigma|}{|\Sigma_{22}|} \qquad \text{by Equation (1.4)}$$

Since $|\Sigma_{22}|$ is the cofactor of σ_{11} in $|\Sigma|$, it follows that

$$\text{Var}(X_1|\mathbf{X}_2 = \mathbf{x}_2) = \frac{1}{\sigma^{11}} \tag{6.8}$$

where σ^{11} is the leading element in Σ^{-1} (if Σ^{-1} exists) and is a constant not dependent on $\mathbf{x}_2$.

The matrix $\Sigma_{12}\Sigma_{22}^{-1}$ is a $q \times (p-q)$ matrix of regression coefficients, the jth row ($j \leq q$) consisting of the regression coefficients for the regression of X_j on

$X_{q+1}, X_{q+2}, \ldots, X_p$. If we apply these results to the bivariate normal distribution discussed in Section 2.4 using the form of Σ in Equation (2.15), we obtain the usual regression line

$$E(X_1|X_2 = x_2) = \mu_1 + \rho\frac{\sigma_1}{\sigma_2}(x_2 - \mu_2)$$

with residual variance

$$\text{Var}(X_1|X_2 = x_2) = \sigma_1^2(1 - \rho^2)$$

The above discussion assumes that Σ_{22} is of full rank. However, the result holds even when Σ_{22}^{-1} does not exist if, in the expressions for the conditional mean and variance, Σ_{22}^{-1} is replaced by Σ_{22}^{-}, a generalized inverse. See Rao (1973, Section 8a2).

We complete this section by noting three results essentially concerned with the algebra of expected values of *linear combinations* of vector random variables.

(v) Suppose $\mathbf{V} = \sum_{r=1}^{n} d_r\mathbf{X}_r$, where the $\mathbf{X}_r$ are p-dimensional random variables and the d_r are scalar constants; then

$$E(\mathbf{V}) = \sum_{r=1}^{n} d_r E(\mathbf{X}_r) \tag{6.9}$$

$$\text{Var}(\mathbf{V}) = \sum_{r=1}^{n} d_r^2 \text{Var}(\mathbf{X}_r) + 2\sum_{r<s} d_r d_s \text{Cov}(\mathbf{X}_r, \mathbf{X}_s) \tag{6.10}$$

Moreover, if the $\mathbf{X}_r$ are normally distributed, so is $\mathbf{V}$.
The derivation of the result is left as an exercise to the reader. Remember that $\text{Cov}(\mathbf{X}_r, \mathbf{X}_s) = [\text{Cov}(\mathbf{X}_s, \mathbf{X}_r)]^{\text{T}}$.
When the $\mathbf{X}_r$ are uncorrelated and have the same variance Σ, then

$$\text{Var}(\mathbf{V}) = \left[\sum_{i=1}^{n} d_r^2\right]\Sigma = (\mathbf{d}^{\text{T}}\mathbf{d})\Sigma \tag{6.11}$$

where $\mathbf{d}^{\text{T}} = [d_1, d_2, \ldots, d_n]$.

(vi) If the $\mathbf{X}_r$ are uncorrelated and have the same variance Σ, and if $\mathbf{V} = \sum_{r=1}^{n} d_{r1}\mathbf{X}_r$ and $\mathbf{W} = \sum_{r=1}^{n} d_{r2}\mathbf{X}_r$, then

$$\text{Cov}(\mathbf{V}, \mathbf{W}) = (\mathbf{d}_1^{\text{T}}\mathbf{d}_2)\Sigma \tag{6.12}$$

where $\mathbf{d}_i^{\text{T}} = [d_{1i}, d_{2i}, \ldots, d_{ni}]$.

Moreover, if the $\mathbf{X}_r$ are normally distributed, $\mathbf{V}$ and $\mathbf{W}$ have normal distributions which are independent if $\mathbf{d}_1^{\text{T}}\mathbf{d}_2 = 0$. The derivation of the result is again left to the reader.

(vii) Suppose the $\mathbf{X}_r$ are independently distributed as $\text{N}_p(\boldsymbol{\mu}, \Sigma)$. Then the sample mean of a random sample of n independent observations is given by

$$\bar{\mathbf{X}} = \frac{1}{n}\sum_{r=1}^{n} \mathbf{X}_r$$

and using the results in (v) with $d_r = 1/n$, we find

$$\bar{\mathbf{X}} \sim \mathrm{N}_p\left(\boldsymbol{\mu}, \frac{1}{n}\Sigma\right) \tag{6.13}$$

Note that the covariance matrix of the sample mean is inversely proportional to the sample size, just as in the univariate case where we have $\mathrm{Var}(\bar{X}) = \sigma^2/n$.

The results expressed in (v)–(vii) are crucial to the development of the sampling theory needed for the methodology of the following three chapters. Before proceeding to this development we shall take the opportunity to emphasize the distinction between our usage of the terms 'linear compound' and 'linear combination'.

6.4 Linear compounds and linear combinations

Suppose $X = \begin{bmatrix} \mathbf{X}_1^{\mathrm{T}} \\ \mathbf{X}_2^{\mathrm{T}} \\ \vdots \\ \mathbf{X}_n^{\mathrm{T}} \end{bmatrix}$ and $E(X) = M = \begin{bmatrix} \boldsymbol{\mu}_1^{\mathrm{T}} \\ \boldsymbol{\mu}_2^{\mathrm{T}} \\ \vdots \\ \boldsymbol{\mu}_n^{\mathrm{T}} \end{bmatrix}$

Note that the data matrix of Section 1.2 and later is a realization of X as used in this chapter. However, the sense intended should always be apparent. The assumption that the $\mathbf{X}_r$ are independently distributed has the implication that when the data are collected, the observation made on any individual is independent of the observation made on any other individual.

The term 'linear compound' is used to refer to a transformation applied to a vector random variable to produce a new (univariate) random variable; for example,

$$Z = \mathbf{c}^{\mathrm{T}}\mathbf{X}$$

Several such linear compounds can be used to define a new vector random variable. Thus

$$\underset{(q\times 1)}{\mathbf{Z}} = \begin{bmatrix} \mathbf{c}_1^{\mathrm{T}} \\ \mathbf{c}_2^{\mathrm{T}} \\ \vdots \\ \mathbf{c}_q^{\mathrm{T}} \end{bmatrix} \mathbf{X} = \underset{(q\times p)}{C^{\mathrm{T}}}\ \underset{(p\times 1)}{\mathbf{X}}$$

The data matrix for a set of n observations on this q-dimensional random variable may be written in terms of the original data matrix X, as

$$\underset{(n\times q)}{Z} = \underset{(n\times p)(p\times q)}{X \quad C} = X[\mathbf{c}_1, \mathbf{c}_2, \ldots, \mathbf{c}_q] = \begin{bmatrix} \mathbf{Z}_1^{\mathrm{T}} \\ \mathbf{Z}_2^{\mathrm{T}} \\ \vdots \\ \mathbf{Z}_n^{\mathrm{T}} \end{bmatrix}$$

Note that we still have the original n observations but that these are now observations on a q-dimensional random variable. If the $\mathbf{X}_r$, $r = 1, 2, \ldots, n$, are independently distributed then so are the $\mathbf{Z}_r$.

When $\mathbf{X}_r \sim \mathrm{N}_p(\boldsymbol{\mu}_r, \Sigma)$ then $\mathbf{Z}_r \sim \mathrm{N}_q(C^{\mathrm{T}}\boldsymbol{\mu}_r, C^{\mathrm{T}}\Sigma\, C)$. Remember also that it is possible to find C such that $C^{\mathrm{T}}\Sigma\, C = I$ (see the discussion of Theorem 6.1).

The term 'linear combination' is used to refer to a weighted sum of vector random variables which is itself a vector random variable, and it will be applied in later sections to linear combinations of the vector responses of different individuals; for example,

$$\mathbf{V}_s = \sum_{r=1}^{n} d_{rs}\mathbf{X}_r = [\mathbf{X}_1, \mathbf{X}_2, \ldots \mathbf{X}_n]\mathbf{d}_s$$

If we define m of these linear combinations (where usually $m < n$), we are in effect defining a set of m new vector responses which have the same number of components measured in the same units as the original data. For example, if $\mathbf{X}_1, \ldots, \mathbf{X}_{n_1}$ are observations on n_1 boys, while $\mathbf{X}_{n_1+1}, \ldots, \mathbf{X}_n$ are observations on $(n - n_1)$ girls, then we can compare the mean vectors for the boys and girls by calculating

$$\mathbf{V}_1 = \sum_{r=1}^{n_1} \mathbf{X}_r/n_1$$

and

$$\mathbf{V}_2 = \sum_{r=n_1+1}^{n} \mathbf{X}_r/(n - n_1)$$

The data matrix for the set of m new vector responses may be written in terms of the original data matrix X, as

$$\underset{(m\times p)}{V} = \underset{(m\times n)(n\times p)}{D^{\mathrm{T}} \quad X} = \begin{bmatrix} \mathbf{d}_1^{\mathrm{T}} \\ \mathbf{d}_2^{\mathrm{T}} \\ \vdots \\ \mathbf{d}_m^{\mathrm{T}} \end{bmatrix} X = \begin{bmatrix} \mathbf{V}_1^{\mathrm{T}} \\ \vdots \\ \mathbf{V}_m^{\mathrm{T}} \end{bmatrix}$$

When the $\mathbf{X}_r$ are independently $\mathrm{N}_p(\boldsymbol{\mu}_r, \Sigma)$, then, from Section 6.3 (vi), the $\mathbf{V}_s$ are independently (and normally) distributed if and only if the $\mathbf{d}_s$ are chosen such that $\mathbf{d}_s^{\mathrm{T}}\mathbf{d}_t = 0$ for all $s \neq t$; $s, t = 1, 2, \ldots, m$. An important special case occurs when D is orthogonal, so that

$$\mathbf{d}_s^{\mathrm{T}}\mathbf{d}_t = \begin{cases} 1 & s = t \\ 0 & s \neq t \end{cases}$$

Then, using Equations (6.11) and (6.12), if the $\mathbf{X}_r$ are independently $N_p(\boldsymbol{\mu}_r, \Sigma)$, the $\mathbf{V}_s$ are independently $N_p(\boldsymbol{v}_s, \Sigma)$, where $\boldsymbol{v}_s = M^T\mathbf{d}_s$.

6.5 Estimation of the parameters of the distribution

It was stated in Chapter 3 that, given a random sample of n independent observations on the vector random variable $\mathbf{X}$, the sample mean vector $\bar{\mathbf{X}}$ and the sample covariance matrix S (with divisor $(n-1)$) were unbiased estimates of $\boldsymbol{\mu}$ and Σ. When $\mathbf{X}$ has a multivariate normal distribution, it can be shown that $\bar{\mathbf{X}}$ is the maximum-likelihood estimator of $\boldsymbol{\mu}$ and that, analogous to the univariate case, $[(n-1)/n]S$ is the maximum-likelihood estimator of Σ and is biased. The derivation of these results is not very enlightening and is omitted. It may be found in Anderson (1958) or Rao (1973). We shall use the unbiased estimators throughout this book. Their joint sampling distribution is derived in Section 6.7 and this will require a study of the Wishart distribution in Section 6.6.

Before leaving this section, however, it should be observed that there are objections to the use of $\bar{\mathbf{X}}$. In 1956, C. Stein showed that when $\mathbf{X}$ has three or more components, $\bar{\mathbf{X}}$ is inadmissible as an estimate of $\boldsymbol{\mu}$, in that we can find an alternative estimator whose expected mean-square error, summed over all components, is less than that of $\bar{\mathbf{X}}$ whatever the value of $\boldsymbol{\mu}$.

The result, an application of decision theory, created a furore in the statistical world and has been followed by the publication of many theoretical papers in decision theory but few on practical applications.

The published papers are concerned with the estimation of $\boldsymbol{\mu}$ when $\mathbf{X} \sim N_p(\boldsymbol{\mu}, \sigma^2 I)$, since it is possible to reduce more general problems to this special case (see, for example, Theorem 6.1), and mainly deal with the case when σ^2 is known. Suppose a random sample of n independent observations is taken from the distribution of $\mathbf{X}$. The maximum-likelihood estimator of $\boldsymbol{\mu}$ is $\bar{\mathbf{X}}$, where $\bar{\mathbf{X}}^T = [\bar{X}_{.1}, \bar{X}_{.2}, \ldots, \bar{X}_{.p}]$, and we note that $\hat{\mu}_j = \bar{X}_{.j}$ is a linear estimator and only involves observations made on the jth component. We are not surprised at this since the X_j, $j = 1, 2, \ldots, p$, are independently distributed.

However, the estimator proposed by James and Stein (1961) is

$$\mu_j^* = \bar{X}_{.j} - \lambda(\bar{X}_{.j} - \mu_0)$$

where $\lambda = (p-2)\sigma^2 \Big/ \left\{ n \sum_{j=1}^{p} (\bar{X}_{.j} - \mu_0)^2 \right\}$ and μ_0 is some chosen value, possibly zero or possibly an initial guess at the mean of the μ_j.

We note that the estimator is non-linear in the observations and involves observations on random variables X_i, $i \neq j$, which are distributed independently of X_j. The effect of the use of this estimator is to 'shrink' the estimates of the μ_j towards the chosen value μ_0.

Essentially the same estimator can be derived by Bayesian methods (see Lindley's contribution to the discussion of Stein, 1962; and Lindley and Smith, 1972), in which the μ_j are considered to be distributed about a prior mean μ_0.

However, the Bayesian argument is not essential, so that application is not confined to problems where a prior distribution can reasonably be considered to exist.

An alternative form of the estimator replaces μ_0 by the mean of the component means $\bar{X}_{..} = \left[\sum_{j=1}^{p} \bar{X}_{.j}\right] \Big/ p$ and the modified estimator is

$$\mu_j^{**} = \bar{X}_{.j} - \lambda'(\bar{X}_{.j} - \bar{X}_{..})$$

where $\lambda' = (p-3)\sigma^2 \Big/ \left\{ n \sum_{j=1}^{p} (\bar{X}_{.j} - \bar{X}_{..})^2 \right\}$ and the shrinking is towards $\bar{X}_{..}$.

The interested reader should consult the papers by Efron and Morris (1973a, 1973b and 1975). The introductory sections in these papers provide a simple survey of the approach and the 1975 paper presents several applications of the use of the James–Stein estimator. The discussion following the 1973a paper and following Stein (1962) and Lindley and Smith (1972) makes fascinating reading.

The general feeling seems to be that applications should be carefully chosen and perhaps should be restricted to problems where the μ_j are somehow 'connected'. There is also a concern that although the Stein-type estimators achieve uniformly lower aggregate risks over all the components, they also permit considerably increased risk to individual components.

It is our impression that the straightforward maximum-likelihood estimator is still used in the vast majority of applications and that this state of affairs will continue unless the Stein approach receives a new impetus through the publication of more practical applications.

6.6 The Wishart distribution

The Wishart distribution is the multivariate generalization of the χ^2 distribution. $\chi^2(f)$ may be defined as the distribution of the sum of squares of f independent standard normal variates. We shall define the Wishart distribution in a similar way and then derive some of its properties.

Definition: Suppose $\mathbf{X}_r$, $r = 1, 2, \ldots, f$, *are independently distributed as* $N_p(\boldsymbol{\mu}_r, \Sigma)$. *Then* $W = \sum_{r=1}^{f} \mathbf{X}_r \mathbf{X}_r^{\mathrm{T}}$ *is said to have a Wishart distribution on* f *degrees of freedom and will be called a Wishart matrix. The distribution is said to be central if* $\boldsymbol{\mu}_r = \mathbf{0}$ *for all* r, *and we write:*

$$W \sim W_p(f, \Sigma)$$

Otherwise it is said to be non-central and we write:

$$W \sim W_p(f, \Sigma; M)$$

where $M^{\mathrm{T}} = [\boldsymbol{\mu}_1, \boldsymbol{\mu}_2, \ldots, \boldsymbol{\mu}_f]$. □

When $p = 1$, it is clear from the above definition that the distribution of $W_1(f, \sigma^2)$ is simply that of $\sigma^2\chi^2(f)$.

Now consider the random matrix

$$X = \begin{bmatrix} \mathbf{X}_1^{\mathrm{T}} \\ \mathbf{X}_2^{\mathrm{T}} \\ \vdots \\ \mathbf{X}_f^{\mathrm{T}} \end{bmatrix}$$

which is of order $(f \times p)$ rather than the more usual $(n \times p)$. X may be written in terms of its column vectors as

$$X = [\mathbf{Y}_1, \mathbf{Y}_2, \ldots, \mathbf{Y}_p]$$

where the $\mathbf{Y}_j$ are $(f \times 1)$ random vectors representing particular components of the random vector $\mathbf{X}$.

Then $$W = X^{\mathrm{T}}X = \sum_{r=1}^{f} \mathbf{X}_r\mathbf{X}_r^{\mathrm{T}} = ((w_{ij})) \tag{6.14}$$

where $w_{ij} = \mathbf{Y}_i^{\mathrm{T}}\mathbf{Y}_j$ is the (ij)th element of W and we may think of W as the random matrix whose elements are the (uncorrected) sums of squares and products of the $\mathbf{Y}_j$.

Let us now consider some of the properties of the distribution.

(i) $$E[W] = f\Sigma + M^{\mathrm{T}}M \tag{6.15}$$

By Equation (2.6), $E[\mathbf{X}_r\mathbf{X}_r^{\mathrm{T}}] = \Sigma + \boldsymbol{\mu}_r\boldsymbol{\mu}_r^{\mathrm{T}}$. The result then follows from the definition of the distribution.

(ii) $\mathrm{Rank}\,(W) = \min\,(f, p)$ with probability 1. (6.16)

When $f < p$, the $\mathbf{X}_r$ are linearly independent vectors with probability 1 and rank $(W) = \mathrm{rank}\,(X^{\mathrm{T}}) = f$. When $f \geq p$, the $\mathbf{X}_r$ span the Euclidean vector space of dimension p with probability 1. Hence X^{T} is of full rank p, so that rank $(W) = p$.

(iii) If $W_1 \sim W_p(f_1, \Sigma; M_1)$ and $W_2 \sim W_p(f_2, \Sigma; M_2)$ independently, then

$$W_1 + W_2 \sim W_p(f_1 + f_2, \Sigma; M) \tag{6.17}$$

where $M^{\mathrm{T}} = [M_1^{\mathrm{T}} \mid M_2^{\mathrm{T}}]$

This result, which may be derived directly from the definition, is a generalization of the additive property of the χ^2 distribution. Note the requirement that the distributions of W_1 and W_2 depend on the same covariance matrix Σ.

(iv) If $W \sim W_p(f, \Sigma; M)$ and C is any $(p \times q)$ matrix of constants, then

$$C^{\mathrm{T}}WC \sim W_q(f, C^{\mathrm{T}}\Sigma C; MC) \tag{6.18}$$

We may write $C^{\mathrm{T}}WC = \sum_{r=1}^{f} C^{\mathrm{T}}\mathbf{X}_r\mathbf{X}_r^{\mathrm{T}}C = \sum_{r=1}^{f} \mathbf{Z}_r\mathbf{Z}_r^{\mathrm{T}}$, say, and the $\mathbf{Z}_r$ are independently $\mathrm{N}_q(C^{\mathrm{T}}\boldsymbol{\mu}_r, C^{\mathrm{T}}\Sigma C)$ (see Section 6.4). The result follows from the definition.

(v) If $W \sim W_p(f, \Sigma; M)$ and $\mathbf{c}$ is any $(p \times 1)$ vector of constants, then

$$\mathbf{c}^T W\mathbf{c} \sim \sigma^2\chi^2(f; \delta^2) \tag{6.19}$$

where $\sigma^2 = \mathbf{c}^T\Sigma\mathbf{c}$ and $\sigma^2\delta^2 = \mathbf{c}^T M^T M\mathbf{c}$.

We may write $W = \sum_{r=1}^{f} \mathbf{X}_r\mathbf{X}_r^T$, where the $\mathbf{X}_r$ are independently $N_p(\boldsymbol{\mu}_r, \Sigma)$. Hence, $\mathbf{c}^T W\mathbf{c} = \sum_{r=1}^{f} (\mathbf{c}^T\mathbf{X}_r)^2$, where the $\mathbf{c}^T\mathbf{X}_r$ are independently $N_1(\mathbf{c}^T\boldsymbol{\mu}_r, \mathbf{c}^T\Sigma\mathbf{c})$ and the distributional result follows. The non-centrality parameter δ^2 is given by $\sigma^2\delta^2 = \sum_{r=1}^{f} (\mathbf{c}^T\boldsymbol{\mu}_r) = \mathbf{c}^T\left(\sum_{r=1}^{f} \boldsymbol{\mu}_r\boldsymbol{\mu}_r^T\right)\mathbf{c} = \mathbf{c}^T M^T M\mathbf{c}$.

Note that if the Wishart distribution is central, so is the χ^2 distribution.

Properties (iv) and (v) are concerned with linear compounds involving transformations to new variables; the remaining properties deal with linear combinations of the individuals.

Suppose

$$X = \begin{bmatrix} \mathbf{X}_1^T \\ \mathbf{X}_2^T \\ \vdots \\ \mathbf{X}_n^T \end{bmatrix}$$

where the $\mathbf{X}_r$ are independently $N_p(\boldsymbol{\mu}_r, \Sigma)$, and $E(X^T) = M^T = [\boldsymbol{\mu}_1, \boldsymbol{\mu}_2, \ldots, \boldsymbol{\mu}_n]$. If $D = [\mathbf{d}_1, \mathbf{d}_2, \ldots, \mathbf{d}_n]$ is any orthogonal matrix of order n and $\mathbf{V}_r = X^T\mathbf{d}_r$, $r = 1, 2, \ldots, n$, then, by the discussion in Section 6.4, the $\mathbf{V}_r$ are also independently $N_p(\boldsymbol{\nu}_r, \Sigma)$ where $\boldsymbol{\nu}_r = M^T\mathbf{d}_r$.

We may now state the remaining properties to be considered.

(vi) The Wishart matrix $X^T X$ may be partitioned into the sum of independently distributed Wishart matrices as

$$X^T X = \sum_{k=1}^{t} X^T D_k D_k^T X \tag{6.20}$$

where the D_k are matrices $\left(\text{of order } (n \times n_k), \text{ say, with } \sum_{k=1}^{t} n_k = n\right)$ whose columns are disjoint subsets of the columns of any orthogonal matrix D.

To derive this result, note that we may write

$$X^T X = X^T D D^T X = \sum_{r=1}^{n} (X^T\mathbf{d}_r)(X^T\mathbf{d}_r)^T = \sum_{r=1}^{n} \mathbf{V}_r\mathbf{V}_r^T$$

where the $\mathbf{V}_r$ are independently $N_p(\boldsymbol{\nu}_r, \Sigma)$ (see Section 6.4).
If $D_k = [\mathbf{d}_{(1)}\mathbf{d}_{(2)}, \ldots, \mathbf{d}_{(n_k)}]$ then, similarly

$$X^T D_k D_k^T X = \sum_{s=1}^{n_k} \mathbf{V}_{(s)}\mathbf{V}_{(s)}^T$$

where the $\mathbf{V}_{(s)}$ form a subset of the $\mathbf{V}_r$. The t matrices $X^T D_k D_k^T X$ are independently distributed since they can be expressed in terms of disjoint subsets of the $\mathbf{V}_r$ and each, by definition is a Wishart matrix on n_k degrees of freedom. Now,

$$\begin{aligned} X^T X &= X^T D D^T X \qquad \text{(since } D \text{ is orthogonal so that } DD^T = I) \\ &= X^T(D_1, D_2, \ldots, D_t)(D_1, D_2, \ldots, D_t)^T X \\ &= \sum_{k=1}^{t} X^T D_k D_k^T X \end{aligned}$$

and the result is established.

When $p = 1$, Equation (6.20) is an expression of the partition of a total sum of squares into quadratic forms independently distributed as $\sigma^2 \chi^2$ distributions, which is the basis of the distribution theory of univariate linear models.

More generally, if $\mathbf{c}$ is any $(p \times 1)$ vector of constants, the linear compounds $Z_r = \mathbf{c}^T \mathbf{X}_r$, $r = 1, 2, \ldots, n$ are distributed independently as $N_1(\mathbf{c}^T \boldsymbol{\mu}_r, \sigma^2)$, where $\sigma^2 = \mathbf{c}^T \Sigma \mathbf{c}$, so that, writing $\mathbf{Z}^T = [Z_1, Z_2, \ldots, Z_n] = \mathbf{c}^T X^T$ and using Equations (6.19) and (6.20) we can partition the sum of squares of the Z_r as follows:

$$\sum_{r=1}^{n} Z_r^2 = \mathbf{Z}^T\mathbf{Z} = \mathbf{c}^T X^T X \mathbf{c} = \sum_{k=1}^{t} \mathbf{c}^T X^T D_k D_k^T X \mathbf{c} = \sum_{k=1}^{t} \mathbf{Z}^T D_k D_k^T \mathbf{Z} \qquad (6.21)$$

where the $\mathbf{Z}^T D_k D_k^T \mathbf{Z}$ are independently distributed as $\sigma^2 \chi^2$. Conversely, it can be shown that, if a sum of squares can be partitioned into a sum of independently distributed $\sigma^2 \chi^2$ variates, the partition must take the form of Equation (6.21). (See, for example, Rao, 1973, Section 3b.4, noting that the $(D_k D_k^T)$ are idempotent.) The same argument (and the same choice of D) justifies the necessity of the form of Equation (6.20).

The theory outlined above is basic to the methodology of the multivariate analysis of variance presented in Chapter 8. Suppose an experiment gives rise to p-variate data, assumed to be normally distributed, and consider performing a univariate analysis of variance on a linear compound of the vector response. The univariate theory allows us to assume the partition of the total sum of squares in Equation (6.21) and the existence of D. Given D we can then write the partition in Equation (6.20). Furthermore, since D is the same matrix for the univariate and the multivariate partitions, it is clear that the diagonal terms of the matrices in Equations (6.20) are simply the sums of squares in the corresponding analysis of variance for the individual components. Calculation of the off-diagonal terms may also be related to the calculation of univariate analyses of variance (see Exercise 6.7).

6.7 The joint distribution of the sample mean vector and the sample covariance matrix

In this section it will be shown that the sample mean vector $\bar{\mathbf{X}}$ and the sample covariance matrix S are independently distributed, analagous to the similar result in univariate theory.

Suppose $\mathbf{X}_r$, $r = 1, 2, \ldots, n$, are identically and independently distributed as $N_p(\boldsymbol{\mu}, \Sigma)$ and that $X^T = [\mathbf{X}_1, \mathbf{X}_2, \ldots, \mathbf{X}_n]$. Let D be an orthogonal matrix of order n whose first column is $\mathbf{d}_1$, where $\mathbf{d}_1^T = (1/\sqrt{n})[1, 1, \ldots 1]$ and partitioned as $D = [\mathbf{d}_1, D_2]$.

Then by Equation (6.20),

$$X^T X = X^T \mathbf{d}_1 \mathbf{d}_1^T X + X^T D_2 D_2^T X \tag{6.22}$$

is a partition of $X^T X$ into two independently distributed Wishart matrices. By Section 6.4 the random vectors $X^T \mathbf{d}_r = \mathbf{V}_r$, say, are independently and normally distributed with covariance matrix Σ.

Now $\mathbf{V}_1 = \frac{1}{\sqrt{n}} \sum_{r=1}^{n} \mathbf{X}_r = \sqrt{n}\bar{\mathbf{X}}$, so that

$$X^T \mathbf{d}_1 \mathbf{d}_1^T X = \mathbf{V}_1 \mathbf{V}_1^T = n\bar{\mathbf{X}}\bar{\mathbf{X}}^T \tag{6.23}$$

$$\begin{aligned} X^T D_2 D_2^T X &= \sum_{r=2}^{n} \mathbf{V}_r \mathbf{V}_r^T \\ &= X^T X - n\bar{\mathbf{X}}\bar{\mathbf{X}}^T \qquad \text{by Equations (6.22) and (6.23)} \\ &= (n-1)S \qquad \text{by Equation (3.5) applied to random variables.} \end{aligned}$$

Then since $\mathbf{V}_1$ is distributed independently of $\mathbf{V}_r$, $r = 2, 3, \ldots, n$, it follows that $\bar{\mathbf{X}}$ and S are independently distributed.

The distribution of $\bar{\mathbf{X}}$ is that of $(1/\sqrt{n})\mathbf{V}_1$, and hence is $N_p(\boldsymbol{\mu}, (1/n)\Sigma)$, a result already derived in Equation (6.13).

To obtain the distribution of $(n-1)S$, note that for $r > 1$ $E[\mathbf{V}_r] = E\left[\sum_{s=1}^{n} d_{sr}\mathbf{X}_r\right] = \left(\sum_{s=1}^{n} d_{sr}\right)\boldsymbol{\mu} = \mathbf{0}$, since the orthogonality of D implies that

$$\mathbf{d}_1^T \mathbf{d}_r = \frac{1}{\sqrt{n}} \sum_{s=1}^{n} d_{sr} = 0$$

From the definition of a Wishart distribution it follows that

$$(n-1)S \sim W_p(n-1, \Sigma) \tag{6.24}$$

and using Equation (6.15),

$$E[S] = \Sigma$$

When $p = 1$, the distributional results reduce to the usual univariate normal theory that the sample mean aand sample variance are independently distributed with $\bar{X} \sim N(\mu, \sigma^2/n)$ and $(n-1)s^2 \sim \sigma^2\chi^2(n-1)$, where s^2 is the unbiased estimator of σ^2.

6.8 The Hotelling T^2-distribution

This section is concerned with the Hotelling T^2-distribution. The distribution is named after Harold Hotelling, who proposed it as a multivariate generalization of the Student t-distribution. The general result which we shall derive may be stated as follows:

Suppose $\mathbf{X} \sim N_p(\boldsymbol{\mu}, (1/k)\Sigma)$ where k is some scalar constant whose value will depend on the application and suppose $fS \sim W_p(f, \Sigma)$ independently, with $f > p-1$, then T^2 is a (univariate) random variable given by

$$T_p^2(f) = k(\mathbf{X} - \boldsymbol{\mu}_0)^{\mathrm{T}} S^{-1}(\mathbf{X} - \boldsymbol{\mu}_0) \tag{6.25}$$

and

$$\frac{f-p+1}{pf} T_p^2(f) \sim F(p, f-p+1; \delta^2) \tag{6.26}$$

a non-central F distribution with non-centrality parameter δ^2 given by

$$\delta^2 = k(\boldsymbol{\mu} - \boldsymbol{\mu}_0)^{\mathrm{T}} \Sigma^{-1}(\boldsymbol{\mu} - \boldsymbol{\mu}_0) \tag{6.27}$$

Note that when $\boldsymbol{\mu} = \boldsymbol{\mu}_0$ the distribution is central, and we write

$$\frac{f-p+1}{pf} T_p^2(f) \sim F(p, f-p+1) \tag{6.28}$$

Hotelling's choice of T^2 was partly motivated by its invariance property; for if $\mathbf{Z} = C^{\mathrm{T}}\mathbf{X} + \mathbf{b}$ where C is a non-singular matrix of constants and $\mathbf{b}$ is a vector of constants, then $S_Z = C^{\mathrm{T}}SC$ and, if we define $\boldsymbol{\xi}_0 = C^{\mathrm{T}}\boldsymbol{\mu}_0 + \mathbf{b}$, we have

$$\begin{aligned} [T_p^2(f)]_Z &= k(\mathbf{Z} - \boldsymbol{\xi}_0)^{\mathrm{T}} S_Z^{-1}(\mathbf{Z} - \boldsymbol{\xi}_0) \\ &= k[C^{\mathrm{T}}(\mathbf{X} - \boldsymbol{\mu}_0)]^{\mathrm{T}}[C^{\mathrm{T}}SC]^{-1}[C^{\mathrm{T}}(\mathbf{X} - \boldsymbol{\mu}_0)] \\ &= [T_p^2(f)]_X \end{aligned} \tag{6.29}$$

so that T_p^2 is invariant under choice of the origin and scale of $\mathbf{X}$.

When $p = 1$, Equation (6.25) reduces to $t^2(f)$, the square of the Student t-statistic on f degrees of freedom, and for the central distributions. Equation (6.28) states the standard univariate result that $t^2(f) \sim F(1, f)$. The invariance property also holds for the t-distribution. We are not surprised to learn that T^2 is used in testing hypotheses on multivariate means, the testing procedures forming the subject of the next chapter. The rest of this section is concerned with the derivation of the distributional result, Equation (6.26). This derivation is rather involved and may be omitted on a first reading, but it is instructive and contains some interesting intermediate results. We proceed as follows:

Suppose $\mathbf{V} \sim N_p(\mathbf{0}, \Sigma)$, where $\mathbf{V}^{\mathrm{T}} = [V_1, V_2, \ldots, V_p]$, and consider the conditional distribution of $V_1|\mathbf{V}_2$ where $\mathbf{V}_2^{\mathrm{T}} = [V_2, V_3, \ldots, V_p]$. This is the 'special case' of Section 6.3 (iv), but even more 'special' since here $E[\mathbf{V}] = \mathbf{0}$.

Corresponding to the partitioning of $\mathbf{V}$, we write

$$\Sigma = \begin{bmatrix} \sigma_{11} & \Sigma_{12} \\ \Sigma_{21} & \Sigma_{22} \end{bmatrix}$$

From Equation (6.7) and (6.8) (with $\boldsymbol{\mu} = \mathbf{0}$) we have

$$E(V_1|\mathbf{V}_2 = \mathbf{v}_2) = \beta_2 v_2 + \beta_3 v_3 + \ldots + \beta_p v_p \tag{6.30}$$

$$\mathrm{Var}(V_1 | \mathbf{V}_2 = \mathbf{v}_2) = \frac{1}{\sigma^{11}} \tag{6.31}$$

where σ^{11} is the leading term of Σ^{-1} (assumed to exist).

Equation (6.30) is the model for the regression *through the origin* of V_1 on $V_2, V_3, \ldots, V_p$, and Equation (6.31) is the variance of V_1 about the regression.

Now suppose we have $f > p - 1$ independent observations on $\mathbf{V}$ regarded as a realization of the set of random vectors $\mathbf{V}_r$, $r = 1, 2, \ldots, f$, independently distributed as $N_p(\mathbf{0}, \Sigma)$, and let $V^{\mathrm{T}} = [\mathbf{V}_1, \mathbf{V}_2, \ldots, \mathbf{V}_f]$. Consider the problem of fitting the model (6.30) given V (of order $(f \times p)$) as the data matrix. Since the model is a regression through the origin, the fitting is carried out using the matrix of uncorrected sums of squares and products, W say, partitioned corresponding to the partition of Σ as

$$W = \begin{bmatrix} w_{11} & W_{12} \\ W_{21} & W_{22} \end{bmatrix}$$

Note that W may be written as $\sum_{r=1}^{f} \mathbf{V}_r \mathbf{V}_r^{\mathrm{T}}$ by Equation (6.14), and hence $W \sim W_p(f, \Sigma)$. W^{-1} exists with probability 1 by Equation (6.16), since $f > p - 1$.

The normal equations for the estimates of β_j are given by

$$W_{22}\hat{\boldsymbol{\beta}}_2 = W_{21} \qquad \text{where } \boldsymbol{\beta}_2^{\mathrm{T}} = [\beta_2, \beta_3, \ldots, \beta_p]$$

and the residual sum of squares is $w_{11} - W_{12}W_{22}^{-1}W_{21} = 1/w^{11}$, where w^{11} is the leading term of W^{-1}, by the same argument used in deriving Equation (6.8).

From regression theory we know that, given f observations on a dependent variable (V_1) and $(p-1)$ regressors ($V_2, V_3, \ldots, V_p$), the residual sum of squares for a regression through the origin is distributed as $\sigma^2\chi^2$ on $f - (p - 1)$ degrees of freedom, where σ^2 is the variance about the regression. That is,

$$\frac{1}{w^{11}} \sim \left(\frac{1}{\sigma^{11}}\right)\chi^2(f - p + 1)$$

or

$$\frac{\sigma^{11}}{w^{11}} \sim \chi^2(f - p + 1) \qquad f > p - 1 \tag{6.32}$$

Of course, we could consider the conditional distribution of any component of $\mathbf{V}$, say V_j, given the other components. Hence, we may generalize the above argument to give the result:

$$\frac{\sigma^{jj}}{w^{jj}} \sim \chi^2(f - p + 1) \qquad f > p - 1 \tag{6.33}$$

A further generalization is possible. Suppose $\mathbf{c}$ is any $(p \times 1)$ vector of constants. Let C be a non-singular matrix whose first column is $\mathbf{c}$ and let $B = C^{-1}$.

By Equation (6.18), $BWB^{\mathrm{T}} \sim W_p(f, B\Sigma B^{\mathrm{T}})$. Now $(BWB^{\mathrm{T}})^{-1} = C^{\mathrm{T}}W^{-1}C$ and similarly, $(B\Sigma B^{\mathrm{T}})^{-1} = C^{\mathrm{T}}\Sigma^{-1}C$. The leading terms in these matrices are $\mathbf{c}^{\mathrm{T}}W^{-1}\mathbf{c}$ and $\mathbf{c}^{\mathrm{T}}\Sigma^{-1}\mathbf{c}$, so that, by Equation (6.32),

$$\frac{\mathbf{c}^{\mathrm{T}}\Sigma^{-1}\mathbf{c}}{\mathbf{c}^{\mathrm{T}}W^{-1}\mathbf{c}} \sim \chi^2(f-p+1) \qquad f>p-1 \tag{6.34}$$

The result is true for any $\mathbf{c}$, even when $\mathbf{c}$ is obtained by observing a random vector which is distributed independently of W.

Hence we have shown that if $\mathbf{U}$, say, and W are independently distributed and $W \sim W_p(f, \Sigma)$, then

$$\frac{\mathbf{U}^{\mathrm{T}}\Sigma^{-1}\mathbf{U}}{\mathbf{U}^{\mathrm{T}}W^{-1}\mathbf{U}} \sim \chi^2(f-p+1) \qquad f>p-1 \tag{6.35}$$

independently of $\mathbf{U}$.

We are now able to complete the derivation of Equation (6.26).

Suppose $\mathbf{X} \sim \mathrm{N}_p(\boldsymbol{\mu}, (1/k)\Sigma)$ and $fS \sim W_p(f, \Sigma)$ independently, and note that $[(1/k)\Sigma]^{-1} = k\Sigma^{-1}$ and $(fS)^{-1} = (1/f)S^{-1}$. We have

$$k(\mathbf{X}-\boldsymbol{\mu}_0)^{\mathrm{T}}\Sigma^{-1}(\mathbf{X}-\boldsymbol{\mu}_0) \sim \chi^2(p;\delta^2) \qquad \text{by Equation (6.5)} \tag{6.36}$$

where $\delta^2 = k(\boldsymbol{\mu}-\boldsymbol{\mu}_0)^{\mathrm{T}}\Sigma^{-1}(\boldsymbol{\mu}-\boldsymbol{\mu}_0)$

and by Equation (6.35), with $\mathbf{U} = \mathbf{X} - \boldsymbol{\mu}_0$,

$$\frac{(\mathbf{X}-\boldsymbol{\mu}_0)^{\mathrm{T}}\Sigma^{-1}(\mathbf{X}-\boldsymbol{\mu}_0)}{(1/f)(\mathbf{X}-\boldsymbol{\mu}_0)^{\mathrm{T}}S^{-1}(\mathbf{X}-\boldsymbol{\mu}_0)} \sim \chi^2(f-p+1) \tag{6.37}$$

independently of $\mathbf{X}$.

Taking ratios,

$$\frac{k}{f}(\mathbf{X}-\boldsymbol{\mu}_0)^{\mathrm{T}}S^{-1}(\mathbf{X}-\boldsymbol{\mu}_0) \sim \frac{\chi^2(p;\delta^2)}{\chi^2(f-p+1)} = \frac{p}{f-p+1}F(p, f-p+1;\delta^2)$$

since the χ^2 distributions are independent, and Equation (6.26) is established.

Exercises

6.1 With $\boldsymbol{\mu}^{\mathrm{T}} = [10, 4, 7]$ and: (i) $B = \begin{bmatrix} 2 & 1 & 0 \\ 1 & 1 & -1 \\ 2 & -1 & 3 \end{bmatrix}$; (ii) $B = \begin{bmatrix} 2 & 1 \\ 1 & 1 \\ 2 & -1 \end{bmatrix}$

write down the distribution of $\mathbf{X} = \boldsymbol{\mu} + B\mathbf{U}$ using Theorem 6.1. Verify the ranks of the resulting covariance matrices.

6.2 Given $\mathbf{X} \sim \mathrm{N}_3(\boldsymbol{\mu}, \Sigma)$, where:

(i) $\Sigma = \begin{bmatrix} 9 & -3 & -3 \\ -3 & 5 & 1 \\ -3 & 1 & 5 \end{bmatrix}$ (ii) $\Sigma = \begin{bmatrix} 8 & -4 & -4 \\ -4 & 4 & 0 \\ -4 & 0 & 4 \end{bmatrix}$

use an eigenvalue analysis to find B, where $\mathbf{X} = \boldsymbol{\mu} + B\mathbf{U}$ as in Theorem 6.1. What structural relationship exists among the X_i for (ii)?

6.3 In Exercise 6.2 (i) let $\mathbf{X} - \boldsymbol{\mu} = B\mathbf{Z} = \begin{bmatrix} k_{11} & 0 & 0 \\ k_{21} & k_{22} & 0 \\ k_{31} & k_{32} & k_{33} \end{bmatrix}\mathbf{Z}$ and choose the k_{ij} with $\mathbf{Z} \sim N_3(\mathbf{0}, I)$ so that $\mathbf{X} = \boldsymbol{\mu} + B\mathbf{Z}$. What is the relationship between $\mathbf{Z}$ and the corresponding $\mathbf{U}$ in Exercise 6.2 (i)? What is the result of using this method in Exercise 6.2(ii)?

6.4 For the distributions of $\mathbf{X}$ appearing in Exercises 6.1 and 6.2, write down the marginal distributions of $\begin{bmatrix} X_1 \\ X_2 \end{bmatrix}$ and the conditional distributions of $X_1 \left\| \begin{bmatrix} X_2 \\ X_3 \end{bmatrix} \right.$ and of $\begin{bmatrix} X_1 \\ X_2 \end{bmatrix} \Big\| X_3$. When are the distributions degenerate? Give the regression functions of X_1 on X_2 and X_3 for the full rank cases. Does a regression function of X_1 on X_2 and X_3 exist in the two cases where rank $(\Sigma) = 2$?

6.5 If $\mathbf{X} \sim N_3(\boldsymbol{\mu}, \Sigma)$ and $C^{\mathrm{T}} = \begin{bmatrix} 1 & -1 & 0 \\ 0 & 1 & -1 \end{bmatrix}$, describe in words what is being measured by the random variable $C^{\mathrm{T}}\mathbf{X}$. What would be the implication if $E(C^{\mathrm{T}}\mathbf{X}) = \mathbf{0}$? If $Z = [1\ \ 1\ \ 1]\mathbf{X}$, find $\operatorname{Cov}(Z, C^{\mathrm{T}}\mathbf{X})$. If C^{T} is replaced by $A^{\mathrm{T}} = \begin{bmatrix} 1 & -1 & 0 \\ 1 & 0 & -1 \end{bmatrix}$, how does this affect the answers to the above question?

6.6 Given $X^{\mathrm{T}} = [\mathbf{X}_1, \mathbf{X}_2, \ldots, \mathbf{X}_6]$ where the $\mathbf{X}_r \sim N_3(\boldsymbol{\mu}_r, \Sigma)$, consider the transformation $V = D^{\mathrm{T}}X$, where

$$D = [\mathbf{d}_1, \mathbf{d}_2, \ldots, \mathbf{d}_6] = \begin{bmatrix} k_1 & k_2 & k_3 & k_4 & 0 & 0 \\ k_1 & k_2 & -k_3 & k_4 & 0 & 0 \\ k_1 & k_2 & 0 & -2k_4 & 0 & 0 \\ k_1 & -k_2 & 0 & 0 & k_5 & k_6 \\ k_1 & -k_2 & 0 & 0 & -k_5 & k_6 \\ k_1 & -k_2 & 0 & 0 & 0 & -2k_6 \end{bmatrix}$$

and the k_r are chosen so that $\mathbf{d}_r^{\mathrm{T}}\mathbf{d}_r = 1$.

If $\mathbf{X}_r$, $r = 1, 2, 3$, are measured on one treatment group and $\mathbf{X}_r$, $r = 4, 5, 6$, measured on another, what is being measured by the random variables $X^{\mathrm{T}}\mathbf{d}_r$? How are they distributed? [*Hint*: try the exercise firstly by assuming the $\mathbf{X}_r$ are univariate normals.]

6.7 The transformation in Exercise 6.6 may be used to partition $X^{\mathrm{T}}X$ into three independent Wishart matrices, $X^{\mathrm{T}}\mathbf{d}_1\mathbf{d}_1^{\mathrm{T}}X$, $X^{\mathrm{T}}\mathbf{d}_2\mathbf{d}_2^{\mathrm{T}}X$ and $\sum_{r=3}^{6} X^{\mathrm{T}}\mathbf{d}_r\mathbf{d}_r^{\mathrm{T}}X$. Show that the diagonal terms of these matrices are respectively the correction factor, the sum of squares for treatments and the residual sum of squares for the individual analyses of variance for the components by showing the algebraic equivalence of the formulation in terms of the given $\mathbf{d}_r$ and the expressions for the analysis-of-variance calculations (in terms of treatment totals and the grand total). Write down the equivalent 'analysis of variance' expressions for the off-diagonal terms.

6.8 Suppose $W \sim W_p(f, \Sigma)$ and is partitioned as $W = \begin{bmatrix} W_{11} & W_{12} \\ W_{21} & W_{22} \end{bmatrix}$ where W_{22}^{-1} exists and W_{11} is $(q \times q)$, $q < p$. Show that $(W_{11} - W_{12}W_{22}^{-1}W_{21}) \sim W_q(f - p + q, \Sigma_{11} - \Sigma_{12}\Sigma_{22}^{-1}\Sigma_{21})$.

6.9 Refer to Exercise 6.6 and show that, if we set

$$\mathbf{V}_2 = X^{\mathrm{T}}\mathbf{d}_2 \quad \text{and} \quad 4S = \sum_{r=3}^{6} X^{\mathrm{T}}\mathbf{d}_r\mathbf{d}_r^{\mathrm{T}}X$$

then $\mathbf{V}_2^{\mathrm{T}}S^{-1}\mathbf{V}_2 \sim T_3^2(4)$
and this may be expressed in terms of a central F-variate if $E[\mathbf{X}]$ is the same for both treatments.

CHAPTER SEVEN
Procedures based on normal distribution theory

7.1 Introduction

In this chapter we introduce a number of data-analytic procedures derived from the normal distribution theory of Chapter 6. Like most procedures based on normal theory, they depend on the values of the sample means, variances and covariances, and so have a heuristic attraction independent of the distributional assumptions. Thus, it could be argued that these procedures are useful methods for the analysis of any data, although, of course, significance levels are only valid when the distributional assumptions apply.

7.2 One-sample procedures

Suppose the $(n \times p)$ data matrix X represents a random sample of n independent observations from a distribution assumed to be $N_p(\boldsymbol{\mu}, \Sigma)$, and we wish to test whether or not the data are consistent with a null hypothesis that $\boldsymbol{\mu} = \boldsymbol{\mu}_0$. In the corresponding univariate problem, the usual procedure is to use the t-test. In the multivariate case, we begin by taking a linear compound of $\mathbf{X}$ to convert from multivariate to univariate data.

Denote this linear compound by $Z = \mathbf{a}^T\mathbf{X}$, where $\mathbf{a}$ is a vector of constants to be determined in some optimal way. Then $Z \sim N_1(\mathbf{a}^T\boldsymbol{\mu}, \mathbf{a}^T\Sigma\mathbf{a})$ and the data vector $X\mathbf{a}$ represents a random sample of n independent observations from the distribution of Z, whose sample mean and variance are

$$\bar{z} = \mathbf{a}^T\bar{\mathbf{X}}$$

$$\text{and} \quad s_z^2 = \mathbf{a}^T S\mathbf{a}$$

where $\bar{\mathbf{X}}$ and S are the estimates of the multivariate parameters $\boldsymbol{\mu}$ and Σ presented in Section 6.5.

Under the null hypothesis $\boldsymbol{\mu} = \boldsymbol{\mu}_0$, we have $E(Z) = \mathbf{a}^T\boldsymbol{\mu}_0$ and the t-test statistic for the z-values is

$$\frac{|\mathbf{a}^T(\bar{\mathbf{x}} - \boldsymbol{\mu}_0)|}{\sqrt{\{\mathbf{a}^T S\mathbf{a}/n\}}} \tag{7.1}$$

In the usual two-sided t-test, the null hypothesis is rejected at significance level α if the test statistic is greater than the appropriate percentage point, $t_{\alpha/2}(n-1)$ or, more conveniently, if

$$\frac{n[\mathbf{a}^{\mathrm{T}}(\bar{\mathbf{x}}-\boldsymbol{\mu}_0)]^2}{\mathbf{a}^{\mathrm{T}}S\mathbf{a}} > t^2_{\alpha/2}(n-1) \tag{7.2}$$

The rejection of the null hypothesis implies, in effect, that the information in the sample is not consistent with the pre-supposition that $E(Z)=\mathbf{a}^{\mathrm{T}}\boldsymbol{\mu}_0$. Since $\mathbf{a}$ is a vector of known constants, this is a contradiction of the hypothesis that $\boldsymbol{\mu}=\boldsymbol{\mu}_0$. However, suppose

$$\boldsymbol{\mu}=\begin{bmatrix}2\\1\\4\end{bmatrix}, \qquad \boldsymbol{\mu}_0=\begin{bmatrix}2\\1\\0\end{bmatrix}$$

The multivariate null hypothesis is false, but if $\mathbf{a}^{\mathrm{T}}$ is of the form $[a_1, a_2, 0]$, the univariate null hypothesis is true. From this and similar examples it is clear that the choice of $\mathbf{a}$ is critical.

The obvious solution is to choose $\mathbf{a}$ so that the inconsistency between sample and hypothesis as measured by Equation (7.1) is greatest. The statistic so obtained is denoted by $\mathscr{T}^2$, and is given by

$$\mathscr{T}^2 = \max_{\mathbf{a}}\left(\frac{n[\mathbf{a}^{\mathrm{T}}(\bar{\mathbf{x}}-\boldsymbol{\mu}_0)]^2}{\mathbf{a}^{\mathrm{T}}S\mathbf{a}}\right) \tag{7.3}$$

Note that replacing $\mathbf{a}$ by $k\mathbf{a}$, where k is any scalar constant, does not alter the value of $\mathscr{T}^2$.

It can be shown that the maximum value is achieved where S is of full rank by taking $\mathbf{a} \propto S^{-1}(\bar{\mathbf{x}}-\boldsymbol{\mu}_0)$, when we find

$$\mathscr{T}^2 = n(\bar{\mathbf{x}}-\boldsymbol{\mu}_0)^{\mathrm{T}}S^{-1}(\bar{\mathbf{x}}-\boldsymbol{\mu}_0) \tag{7.4}$$

The result may be derived by differentiation (see Morrison, 1976) or as an application of the Cauchy–Schwartz inequality (see Exercise 7.1). Note that if $p=1$, then Equation (7.4) reduces to the square of the familiar univariate t-statistic. It is often convenient to calculate $\mathscr{T}^2$ by first solving for $\mathbf{a}^* = S^{-1}(\bar{\mathbf{x}}-\boldsymbol{\mu}_0)$ and then calculating $\mathscr{T}^2 = n(\bar{\mathbf{x}}-\boldsymbol{\mu}_0)^{\mathrm{T}}\mathbf{a}^*$, since the value of $\mathbf{a}^*$ is useful in further analysis (see Equation (7.17) below).

The argument above is essentially that given by S. N. Roy in terms of his union–intersection principle. It is presented in detail in Morrison (1976). However, optimal properties of the test seem to depend on the expectation that the properties of the univariate t-test carry through to the multivariate test. It is more satisfactory to observe that $\mathscr{T}^2$ may be derived as the likelihood-ratio test of the hypothesis. The derivation may be found in Anderson (1958) together with a discussion of the optimal properties.

As $\mathscr{T}^2$ is obtained by maximizing over $\mathbf{a}$, its sampling distribution is no longer a t^2-distribution and we must find this distribution in order to carry out tests of hypotheses.

To obtain the sampling distribution of $\mathcal{T}^2$, note that by the results of Section 6.7,

$$\bar{\mathbf{X}} \sim N_p\left(\boldsymbol{\mu}, \frac{1}{n}\Sigma\right) \quad \text{and} \quad (n-1)S \sim W_p(n-1, \Sigma) \tag{7.5}$$

independently.
Hence, when $n > p$,

$$\mathcal{T}^2 = n(\bar{\mathbf{X}} - \boldsymbol{\mu}_0)^{\mathrm{T}} S^{-1}(\bar{\mathbf{X}} - \boldsymbol{\mu}_0) \tag{7.6}$$

is of the form of Equation (6.25) and is distributed as $T_p^2(n-1; \delta^2)$, where

$$\delta^2 = n(\boldsymbol{\mu} - \boldsymbol{\mu}_0)^{\mathrm{T}} \Sigma^{-1}(\boldsymbol{\mu} - \boldsymbol{\mu}_0) \tag{7.7}$$

Under H_0: $\boldsymbol{\mu} = \boldsymbol{\mu}_0$, $\delta^2 = 0$, so that

$$\mathcal{T}^2 \sim T_p^2(n-1) \tag{7.8}$$

and, by Equation (6.26),

$$\frac{(n-p)\mathcal{T}^2}{p(n-1)} \sim F(p, n-p) \tag{7.9}$$

Consequently, the null hypothesis is rejected at significance level α if

$$\frac{(n-p)\mathcal{T}^2}{p(n-1)} > F_\alpha(p, n-p) \tag{7.10}$$

The test is known as the Hotelling T^2-test.

Example 7.1. The following artificial data refers to a sample of approximately two-year-old boys from a high-altitude region in Asia. MUAC is the mid-upper-arm circumference.

For lowland children of the same age in the same country, the height, chest and MUAC means are considered to be 90, 58 and 16 cm respectively. Test

Table 7.1

Individual	Height (cm)	Chest circumference (cm)	MUAC (cm)
1	78	60.6	16.5
2	76	58.1	12.5
3	92	63.2	14.5
4	81	59.0	14.0
5	81	60.8	15.5
6	84	59.5	14.0

the hypothesis that the highland children have the same means.

$$H_0: \boldsymbol{\mu}_0^{\mathrm{T}} = [90, 58, 16]$$

$$\bar{\mathbf{x}}^{\mathrm{T}} = [82.0, 60.2, 14.5] \qquad (\bar{\mathbf{x}} - \boldsymbol{\mu}_0)^{\mathrm{T}} = [-8.0, 2.2, -1.5]$$

$$S = \begin{bmatrix} 31.600 & 8.040 & 0.500 \\ 8.040 & 3.172 & 1.310 \\ 0.500 & 1.310 & 1.900 \end{bmatrix}$$

$$S^{-1} = (23.13848)^{-1} \begin{bmatrix} 4.3107 & -14.6210 & 8.9464 \\ -14.6210 & 59.7900 & -37.3760 \\ 8.9464 & -37.3760 & 35.5936 \end{bmatrix}$$

$$\mathscr{T}^2 = 6 \times 70.0741 = 420.4447$$

$$\mathscr{F} = \frac{n-p}{p(n-1)} \mathscr{T}^2 = 84.09 \qquad F_{.01}(3, 3) = 29.5$$

We conclude that there is strong evidence against the null hypothesis ($P = 0.002$).†

If we use the alternative method of calculating $\mathscr{T}^2$ by solving $S\mathbf{a}^* = (\bar{\mathbf{x}} - \boldsymbol{\mu}_0)$, we obtain

$$\mathbf{a}^{*\mathrm{T}} = [-3.4605, 13.1629, -8.9543]$$

$$(\bar{\mathbf{x}} - \boldsymbol{\mu}_0)^{\mathrm{T}} \mathbf{a}^* = 70.0738$$

and $\mathscr{T}^2 = 420.4430$ which is within rounding error of the earlier result.

It is interesting to note that differentiating $\mathscr{T}^2$ with respect to $(\bar{\mathbf{x}} - \boldsymbol{\mu}_0)$ we obtain

$$\frac{\mathrm{d}\mathscr{T}^2}{\mathrm{d}(\bar{\mathbf{x}} - \boldsymbol{\mu}_0)} = 2nS^{-1}(\bar{\mathbf{x}} - \boldsymbol{\mu}_0) = 2n\mathbf{a}^*$$

so that the elements in $\mathbf{a}^*$ indicate the sensitivity of the value of $\mathscr{T}^2$ to changes in the corresponding elements of $(\bar{\mathbf{x}} - \boldsymbol{\mu}_0)$ for the given units of measurement. Here we see that an increase of 1 cm in the observed mean chest measurement would lead to a greater increase in the value of $\mathscr{T}^2$ than would a 1 cm decrease in either mean height or mean MUAC. It should be emphasized that, although $\mathscr{T}^2$ is invariant to choice of scales of measurement, $\mathbf{a}^*$ is not. For example, if height had been measured in metres, the first element of $(\bar{\mathbf{x}} - \boldsymbol{\mu}_0)$ would be 0.08 and the corresponding element in $\mathbf{a}^*$ would become -346.05, leaving $\mathscr{T}^2$ unaltered. □

7.3 Confidence intervals and further analysis

If in Equation (7.6) we replace $\boldsymbol{\mu}_0$ by the true (but unknown) value $\boldsymbol{\mu}$, then $\mathscr{T}^2 = n(\bar{\mathbf{X}} - \boldsymbol{\mu})^{\mathrm{T}} S^{-1} (\bar{\mathbf{X}} - \boldsymbol{\mu})$ has the central distribution, so that

† The exact significance levels quoted in brackets were obtained using the statistics module programs on a Texas T.I. 58 calculator.

Equation (7.9) applies. Now, since $P(F < F_\alpha) = 1 - \alpha$,

$$P\left[\frac{(n-p)\mathscr{T}^2}{p(n-1)} < F_\alpha(p, n-p)\right] = 1 - \alpha \tag{7.11}$$

Hence

$$P\left[n(\bar{\mathbf{X}} - \boldsymbol{\mu})^{\mathrm{T}} S^{-1}(\bar{\mathbf{X}} - \boldsymbol{\mu}) < \frac{p(n-1)}{n-p} F_\alpha(p, n-p)\right] = 1 - \alpha \tag{7.12}$$

For observed values $\bar{\mathbf{x}}$ and S, we may rewrite the inequality within the brackets as a function of $\boldsymbol{\mu}$ in the form

$$n(\boldsymbol{\mu} - \bar{\mathbf{x}})^{\mathrm{T}} S^{-1}(\boldsymbol{\mu} - \bar{\mathbf{x}}) < \frac{p(n-1)}{n-p} F_\alpha(p, n-p) \tag{7.13}$$

The region defined by Equation (7.13) is a $100(1-\alpha)$ per cent confidence region for $\boldsymbol{\mu}$. The boundary of this region is a hyperellipsoid centred at the point $\boldsymbol{\mu} = \bar{\mathbf{x}}$. Note that the interior points of the region represent those values of $\boldsymbol{\mu}_0$ which would not be rejected in a single application of the Hotelling T^2-test for the given data. However, the calculation involved in finding out whether a given $\boldsymbol{\mu}^*$ lies inside or outside the confidence region is exactly the same as the calculation of $\mathscr{T}^2$ for the test of the hypothesis $H_0: \boldsymbol{\mu} = \boldsymbol{\mu}^*$. Consequently, the technique provides little additional insight except possibly in the bivariate case when the boundary can be plotted (see Example 7.4).

An alternative approach depends on the fact that $\mathscr{T}^2$ is the maximum value of the univariate t^2 for any linear compound of $\mathbf{X}$. The argument is as follows:

$$\mathscr{T}^2 = \max_{\mathbf{a}} \left(\frac{n[\mathbf{a}^{\mathrm{T}}(\bar{\mathbf{X}} - \boldsymbol{\mu})]^2}{\mathbf{a}^{\mathrm{T}} S \mathbf{a}}\right)$$

so that, by Equation (7.11),

$$P\left[\max_{\mathbf{a}} \left(\frac{n[\mathbf{a}^{\mathrm{T}}(\bar{\mathbf{X}} - \boldsymbol{\mu})]^2}{\mathbf{a}^{\mathrm{T}} S \mathbf{a}}\right) < \frac{p(n-1)}{(n-p)} F_\alpha(p, n-p)\right] = 1 - \alpha$$

Hence, simultaneously for all $\mathbf{a}$,

$$P\left[[\mathbf{a}^{\mathrm{T}}(\bar{\mathbf{X}} - \boldsymbol{\mu})]^2 < \frac{p(n-1)}{n-p} F_\alpha(p, n-p) \frac{s_a^2}{n}\right] = 1 - \alpha$$

where $s_a^2 = \mathbf{a}^{\mathrm{T}} S \mathbf{a}$ is the estimate of $\mathrm{Var}(\mathbf{a}^{\mathrm{T}}\mathbf{X})$.
Now the event

$$[\mathbf{a}^{\mathrm{T}}(\bar{\mathbf{X}} - \boldsymbol{\mu})]^2 < \frac{p(n-1)}{n-p} F_\alpha(p, n-p) \frac{s_a^2}{n}$$

implies

$$|\mathbf{a}^{\mathrm{T}}(\bar{\mathbf{X}} - \boldsymbol{\mu})| < \left[\frac{p(n-1)}{n-p} F_\alpha(p, n-p)\right]^{1/2} \frac{s_a}{\sqrt{n}}$$

$$= K_{\alpha/2} \frac{s_a}{\sqrt{n}}, \text{ say} \tag{7.14}$$

or, alternatively, that

$$\mathbf{a}^{\mathrm{T}}\bar{\mathbf{X}} - K_{\alpha/2}\frac{s_a}{\sqrt{n}} < \mathbf{a}^{\mathrm{T}}\boldsymbol{\mu} < \mathbf{a}^{\mathrm{T}}\bar{\mathbf{X}} + K_{\alpha/2}\frac{s_a}{\sqrt{n}}$$

Consequently, for an observed $\bar{\mathbf{x}}$,

$$\mathbf{a}^{\mathrm{T}}\bar{\mathbf{x}} \pm K_{\alpha/2}\frac{s_a}{\sqrt{n}} \qquad \text{for all } \mathbf{a} \tag{7.15}$$

is a set of simultaneous confidence intervals for $\mathbf{a}^{\mathrm{T}}\boldsymbol{\mu}$ with confidence coefficient $100(1-\alpha)$ per cent.

This result may be interpreted in terms of long-run probabilities by saying that of a large number of such *sets* of confidence intervals, we would expect that $100(1-\alpha)$ per cent of *sets* would contain no mis-statements while 100 α per cent of *sets* would contain at least one mis-statement.

This approach, due to Roy and Bose (1953) is essentially the same as that used by Scheffé (1959) in constructing a set of simultaneous confidence intervals for all contrasts in the analysis of variance.

We can in fact rewrite Equation (7.15) as a set of test statistics to test the null hypotheses $H_0(\mathbf{a}):\mathbf{a}^{\mathrm{T}}\boldsymbol{\mu} = \mathbf{a}^{\mathrm{T}}\boldsymbol{\mu}_0$ for different values of $\mathbf{a}$, in the form

$$\frac{\mathbf{a}^{\mathrm{T}}(\bar{\mathbf{x}} - \boldsymbol{\mu}_0)}{s_a/\sqrt{n}} \tag{7.16}$$

rejecting the null hypotheses for those values of $\mathbf{a}$ for which the statistic is greater than $K_{\alpha/2}$.

The tests have the form of the usual two-sided t-test but with a critical value $K_{\alpha/2}$ instead of $t_{\alpha/2}(n-1)$. This ensures that the overall type 1 error is α no matter how many different values of $\mathbf{a}$ are used. Note that $K_{\alpha/2} \geq t_{\alpha/2}(n-1)$, with equality when $p = 1$.

The relationship between the confidence-region approach (Equation (7.11)–(7.13)) and the T^2-test has already been mentioned. A relationship also exists between the T^2-test and the Roy–Bose set of simultaneous confidence intervals. For suppose in an application of the T^2-test, the null hypothesis $H_0:\boldsymbol{\mu} = \boldsymbol{\mu}_0$ is rejected at significance level α, then there exists at least one linear compound such that the interval $\mathbf{a}^{\mathrm{T}}\bar{\mathbf{x}} \pm K_{\alpha/2}\, s_a/\sqrt{n}$ does not include the value $\mathbf{a}^{\mathrm{T}}\boldsymbol{\mu}_0$. This follows immediately from Equation (7.3).

It should be emphasized that the two procedures – the construction of a confidence region or of a set of simultaneous confidence intervals – are not dependent on a prior T^2-test. In fact, the two procedures are estimation procedures although, as we have observed, they implicitly perform tests of hypotheses.

In practice, we are often concerned with identifying in more detail the discrepancies between sample and hypothesis given that we have rejected the null hypothesis. For example, may we infer that, for the ith component of $\mathbf{X}$, $\mu_i \neq \mu_{0i}$? Or may we conclude that $\mu_i - \mu_j = \mu_{0i} - \mu_{0j}$? The obvious approach is to use the simultaneous confidence intervals, Equation (7.15), or the equivalent test statistics, Equation (7.16). However, as in the use of Scheffé's simultaneous confidence intervals in the examination of contrasts following an analysis of variance, the procedure tends to be too demanding in the sense that the proportion of real discrepancies present but not detected (i.e., not found significant) is usually quite large. A less rigorous approach is to compare the test statistics, Equation (7.16), with $t_{\alpha/2}(n-1)$ instead of $K_{\alpha/2}$ (or to construct the equivalent confidence intervals). The overall type 1 error for such a series of tests is not known, since they are performed conditional on the rejection of the null hypothesis by the T^2-test. (This procedure is analogous to the use of the least significant difference after an analysis of variance.) Since $t_{\alpha/2} < K_{\alpha/2}$, the use of the t-value will result in the detection of more real discrepancies, but at the expense of finding more spurious ones.

The problem is essentially a decision problem and could possibly be solved by the application of decision theory if the losses or utilities could be quantified. Failing this, it would seem reasonable to use the t-critical value when the detection of spurious effects is considered a less serious error than the failure to detect real effects; and to use the K-critical value when the contrary situation applies. For any particular linear contrast of the means, a test statistic outside the K-value will certainly be judged significant, while a test statistic inside the t-value will not be judged significant. Values between the t- and K-values must be interpreted by taking account of the consequences of wrong decisions.

In Example 7.1 we rejected the null hypothesis. Let us now examine the data further. The Roy–Bose confidence intervals for the means of the three components with $K_{.025} = \left[\frac{p(n-1)}{n-p}F_{.05}(p, n-p)\right]^{1/2} = (5 \times 9.28)^{1/2} = 6.81$ are:

$$\mu_1 \in 82.0 \pm K_{.025}\sqrt{\frac{31.600}{6}} = 82.0 \pm 15.6$$

$$\mu_2 \in 60.2 \pm K_{.025}\sqrt{\frac{3.172}{6}} = 60.2 \pm 5.0$$

$$\mu_3 \in 14.5 \pm K_{.025}\sqrt{\frac{1.900}{6}} = 14.5 \pm 3.8$$

We note that none of these confidence intervals exclude the original null hypothesis values!

The corresponding 't' confidence intervals, taking $t_{.025}(5) = 2.57$, are

$$\mu_1 \in 82.0 \pm 6.6$$
$$\mu_2 \in 60.2 \pm 1.9$$
$$\mu_3 \in 14.5 \pm 1.6$$

The first two of these confidence intervals exclude the null hypothesis values.

Example 7.2. Before commenting further, let us change the problem slightly.

Suppose the data in Table 7.1 refer to highland children of a particular ethnic stock and that the means for lowland children of the same ethnic stock are given by $\boldsymbol{\mu}_0^{\mathrm{T}} = [88.0, 58.4, 16.0]$. Repeating the calculations of Example 7.1, we obtain

$$(\bar{\mathbf{x}} - \boldsymbol{\mu}_0)^{\mathrm{T}} = [-6.0, 1.8, -1.5]$$

with $\mathscr{T}^2 = 287.2284$

$$\mathscr{F} = \frac{(n-p)}{p(n-1)}\mathscr{T}^2 = 57.45$$

which is again significant at the 1 per cent level ($P = 0.004$).
The confidence intervals for the μ_i are still those calculated in Example 7.1 and we find that *none* of the confidence intervals exclude the null hypothesis values for the component means, not even those calculated using critical values from the t-distribution. Hence we find that, whereas the overall test statistic is highly significant, individual tests on the components do not lead to the rejection of the corresponding null hypothesis. However, consider for example the linear compound $U = x_1 - 2x_2 + x_3$. Under the null hypothesis, $\mu_U = -12.8$. The sample statistics are $\bar{u} = -23.9$ and $s_U^2 = 9.788$, giving for Equation (7.16),

$$\frac{|-23.9 - (-12.8)|}{\sqrt{(9.788/6)}} = 8.69$$

which is significant (at the 5 per cent level) even by the more demanding K-test. □

Example 7.2 is particularly instructive in illustrating the strength of the multivariate approach which uses the information in the sample about correlations among the components. It also underlines the importance of examining linear compounds and of not restricting the further analysis to individual components. One linear compound that should always be examined is that linear compound which demonstrates the greatest discrepancy between sample and hypothesis. This compound is $\mathbf{a}^{*\mathrm{T}}\mathbf{X}$, where (see Equation (7.4))

$$S\mathbf{a}^* = (\bar{\mathbf{x}} - \boldsymbol{\mu}_0) \tag{7.17}$$

From Equation (7.3), the resulting value of the test statistic (7.16) is $\mathscr{T} = t_{\max}$. (Check this by substituting $\mathbf{a} = \mathbf{a}^* = S^{-1}(\bar{\mathbf{x}} - \boldsymbol{\mu}_0)$ in Equation (7.16).)

In Example 7.1, we find that

$$\mathbf{a}^{*\mathrm{T}} = [-3.4605, 13.1629, -8.9543]$$

and the test statistic $t_{\max} = \sqrt{420.4447} = 20.50$

In Example 7.2, $\mathbf{a}^{*\mathrm{T}} = [-2.8352, 10.8655, -7.5349]$ and $t_{\max} = \sqrt{287.2284} = 16.95$

The significance levels using the K-tests are, of course, just the significance levels of the overall Hotelling T^2-test.

Finally, we may sum up (in both examples) by saying that the mean of the sampled population is significantly different from $\boldsymbol{\mu}_0$ and the main discrepancy occurs in a contrast between the chest circumference and the other dimensions.

7.4 Tests of structural relations among the components of the mean

In Section 7.3 the null hypothesis specified the means completely. It is also possible to test hypotheses which specify the existence of relations among the components of the mean vector.

Consider an example where, for each individual, the same variate is measured under p different conditions for instance, at different specified times. An hypothesis of interest is $H_0: \mu_j = \mu; j = 1, 2, \ldots, p$. The hypothesis may be reformulated as $H_0: \mu_1 - \mu_j = 0; j = 2, \ldots, p$, which may be stated in the form:

$$H_0: C^{\mathrm{T}}\boldsymbol{\mu} = \mathbf{0}$$

where
$$C = \begin{bmatrix} 1 & 1 & \ldots & 1 \\ -1 & 0 & \ldots & 0 \\ 0 & -1 & \ldots & 0 \\ \vdots & & & \\ 0 & 0 & & -1 \end{bmatrix} \text{ is of order } (p \times (p-1)) \tag{7.18}$$

Now $C^{\mathrm{T}}\mathbf{X} \sim \mathrm{N}_{p-1}(C^{\mathrm{T}}\boldsymbol{\mu}, C^{\mathrm{T}}\Sigma C)$ and the null hypothesis can be tested by the Hotelling T^2-test using the statistic

$$\mathscr{T}^2 = n\bar{\mathbf{X}}^{\mathrm{T}}C(C^{\mathrm{T}}SC)^{-1}C^{\mathrm{T}}\bar{\mathbf{X}} \tag{7.19}$$

where we require $n > p - 1$.

Under the null hypothesis, $\mathscr{T}^2 \sim T^2_{p-1}(n-1)$, so that

$$\mathscr{F} = \frac{n-p+1}{(p-1)(n-1)}\mathscr{T}^2 \sim F(p-1, n-p+1)$$

Note that C is not unique for testing this hypothesis. For instance,

$$C^* = \begin{bmatrix} 1 & 0 & \ldots & 0 \\ -1 & 1 & \ldots & 0 \\ 0 & -1 & \ldots & 0 \\ \vdots & & & \\ 0 & 0 & & -1 \end{bmatrix}$$

would give the same result. In fact, so would any $(p \times (p-1))$ matrix of the form CA where A is a non-singular $((p-1) \times (p-1))$ matrix. This is an application of the invariance property of $\mathscr{T}^2$. (See Equation (6.29) and Exercise 6.5.)

For a second example, suppose $\mathbf{X}$, with p odd, represents a series of temperature measurements taken at fixed intervals across the back of a patient at the level of a particular vertebra, with the middle observation taken at the centre of the vertebra. We might wish to consider a hypothesis that $\mu_1 = \mu_p, \mu_2 = \mu_{(p-1)}, \ldots$, with an alternative hypothesis indicating a 'hot-spot' which could indicate the presence of cancer. The null hypothesis could be formulated as

$$H_0: C^{\mathrm{T}}\boldsymbol{\mu} = \mathbf{0}$$

where

$$C = \begin{bmatrix} 1 & 0 & \ldots & 0 \\ 0 & 1 & \ldots & 0 \\ & & & \vdots \\ & & & 1 \\ \vdots & & & 0 \\ & & & -1 \\ & & & \vdots \\ 0 & -1 & & 0 \\ -1 & 0 & & 0 \end{bmatrix} \text{ is of order } (p \times \tfrac{1}{2}(p-1)) \tag{7.20}$$

$C^{\mathrm{T}}\mathbf{X} \sim \mathrm{N}_{(p-1)/2}(C^{\mathrm{T}}\boldsymbol{\mu}, C^{\mathrm{T}}\Sigma C)$ and the null hypothesis $H_0: C^{\mathrm{T}}\boldsymbol{\mu} = \mathbf{0}$ is tested by the same statistic as in Equation (7.19) where $n > (p-1)/2$. Under the null hypothesis $\mathcal{T}^2 \sim T^2{}_{(p-1)/2}(n-1)$, so that

$$\frac{n-(p-1)/2}{(p-1)(n-1)/2}\mathcal{T}^2 \sim F((p-1)/2, n-(p-1)/2) \tag{7.21}$$

In general, the statistic for testing the hypothesis $H_0: C^{\mathrm{T}}\boldsymbol{\mu} = \boldsymbol{\phi}$, where $\boldsymbol{\phi}$ is a given vector of constants, is

$$\mathcal{T}^2 = n(C^{\mathrm{T}}\bar{\mathbf{X}} - \boldsymbol{\phi})^{\mathrm{T}}(C^{\mathrm{T}}\Sigma C)^{-1}(C^{\mathrm{T}}\bar{\mathbf{X}} - \boldsymbol{\phi}) \tag{7.22}$$

where C is a given matrix of order $(p \times m)$, say, of rank m, and $n > m$. Under H_0, $\mathcal{T}^2$ is distributed as $T^2_m(n-1)$.

Other structural relations may be defined by expressing the mean as a linear model in terms of $k < p$ parameters. For example, suppose we wish to test the hypothesis $H_0: \boldsymbol{\mu} = A\boldsymbol{\theta}$, where A is a design matrix of known constants, of order $(p \times k)$ and of rank k. If $\boldsymbol{\theta}$ is also fully specified in the null hypothesis, the appropriate test statistic is

$$\mathcal{T}^2 = n(\bar{\mathbf{X}} - A\boldsymbol{\theta})^{\mathrm{T}}S^{-1}(\bar{\mathbf{X}} - A\boldsymbol{\theta}) \tag{7.23}$$

which, under the null hypothesis, is distributed as $T^2_p(n-1)$. Essentially we are testing the hypothesis that the residuals have zero mean.

However, in practice it will usually be necessary to estimate $\boldsymbol{\theta}$. The problem is treated in Morrison (1976, Section 5.7) who uses a technique due to Khatri (1966). Rao (1973) chooses $\hat{\boldsymbol{\theta}}$ to minimize (7.23) and states that the resulting minimum $\mathcal{T}^2$ is distributed as $T^2_{p-k}(n-1)$. (See also Exercise 7.3)

Example 7.3. In Example 7.1 suppose there is an 'anthropometric law' which states that the means of height, chest circumference and MUAC (mid-upper-arm circumference) are in the ratios 6 : 4 : 1 and we wish to test whether or not our data are consistent with this law. The null hypothesis is

$$H_0 : \frac{1}{6}\mu_{10} = \frac{1}{4}\mu_{20} = \mu_{30}$$

which may be reformulated as

$$H_0 : C^{\mathrm{T}}\boldsymbol{\mu}_0 = \mathbf{0} \text{ where } C^{\mathrm{T}} = \begin{bmatrix} 2 & -3 & 0 \\ 1 & 0 & -6 \end{bmatrix}$$

$$C^{\mathrm{T}}\bar{\mathbf{X}} = \begin{bmatrix} -16.6 \\ -4.0 \end{bmatrix} \qquad C^{\mathrm{T}}SC = \begin{bmatrix} 58.468 & 56.660 \\ 56.660 & 94.000 \end{bmatrix}$$

$$\text{hence} \quad (C^{\mathrm{T}}SC)^{-1} = (2285.6364)^{-1}\begin{bmatrix} 94.000 & -56.660 \\ -56.660 & 58.468 \end{bmatrix}$$

and solving $(C^{\mathrm{T}}SC)\mathbf{a}^* = C^{\mathrm{T}}(\bar{\mathbf{x}} - \boldsymbol{\mu}_0)$

we obtain $\mathbf{a}^{*\mathrm{T}} = [-0.5835, 0.3092]$

$$\mathcal{T}^2 = 6 \times 8.450023 = 50.7001 \sim T_2(5) \text{ under } H_0$$

$$\mathcal{F} = 20.28 \qquad F_{.01}(2, 4) = 18.00 \qquad (P = .008)$$

We conclude that the sample is inconsistent with the null hypothesis. The reader could try this example again with

$$C^{\mathrm{T}} = \begin{bmatrix} 0 & 1 & -4 \\ 1 & 0 & -6 \end{bmatrix}$$

□

7.5 Two-sample procedures

Suppose we have collected data from two populations and wish to compare their means. If the samples are of sizes n_1, n_2 respectively, then for $i = 1, 2$, the data matrix X_i is of order $(n_i \times p)$ and represents a random sample of n_i independent observations from a distribution assumed to be $\mathrm{N}_p(\boldsymbol{\mu}_i, \Sigma)$. Note particularly that we assume the covariance matrices of the two populations to be the same. From Section 6.3 (v)–(vii),

$$\bar{\mathbf{X}}_1 - \bar{\mathbf{X}}_2 \sim \mathrm{N}_p\left(\boldsymbol{\mu}_1 - \boldsymbol{\mu}_2, \left[\frac{1}{n_1} + \frac{1}{n_2}\right]\Sigma\right)$$

The pooled within-groups estimate of Σ is given by

$$S = \frac{(n_1 - 1)S_1 + (n_2 - 1)S_2}{n_1 + n_2 - 2} \tag{7.24}$$

Using Equation (6.17) we see that $(n_1 + n_2 - 2)S \sim W_p(n_1 + n_2 - 2, \Sigma)$ and so by Equation (6.25), if $n_1 + n_2 > p + 1$,

$$\mathcal{T}^2 = \frac{n_1 n_2}{n_1 + n_2}(\bar{\mathbf{X}}_1 - \bar{\mathbf{X}}_2)^{\mathrm{T}}S^{-1}(\bar{\mathbf{X}}_1 - \bar{\mathbf{X}}_2) \sim T_p^2(n + n_2 - 2; \delta^2)$$

where
$$\delta^2 = \frac{n_1 n_2}{n_1 + n_2}(\boldsymbol{\mu}_1 - \boldsymbol{\mu}_2)^{\mathrm{T}} \Sigma^{-1} (\boldsymbol{\mu}_1 - \boldsymbol{\mu}_2) \tag{7.25}$$

If $\boldsymbol{\mu}_1 = \boldsymbol{\mu}_2$, $\delta^2 = 0$ and

$$\mathscr{T}^2 \sim T_p^2(n_1 + n_2 - 2)$$

with
$$\frac{n_1 + n_2 - p - 1}{p(n_1 + n_2 - 2)} \mathscr{T}^2 \sim F(p, n_1 + n_2 - p - 1) \tag{7.26}$$

As in the one-sample case, we will make use of the solution of the equation $\mathbf{a}^* = S^{-1}(\bar{\mathbf{x}}_1 - \bar{\mathbf{x}}_2)$ so that $\mathscr{T}^2$ is conveniently calculated as

$$\frac{n_1 n_2}{n_1 + n_2}(\bar{\mathbf{x}}_1 - \bar{\mathbf{x}}_2)^{\mathrm{T}} \mathbf{a}^* \tag{7.27}$$

(see Equation (7.32) in the following section.)

The assumption that the covariance matrices of the two populations are equal is a generalization of the assumption of equal variances in the univariate two-sample t-test. It is possible to test the assumption using a likelihood-ratio test suggested by M. S. Bartlett and modified by Box (1949). However, the $\mathscr{T}^2$-statistic is not sensitive to departures from the assumption when the sample sizes are approximately equal (Ito and Schull, 1964) and is the only statistic in common use.

7.6 Confidence intervals and further analysis

Having rejected the null hypothesis in the overall test, further analysis is again concerned with identifying the discrepancies in more detail.

If $\boldsymbol{\Delta} = \boldsymbol{\mu}_1 - \boldsymbol{\mu}_2$, then analogously to Equation (7.13) the equation

$$\frac{n_1 n_2}{n_1 + n_2}[\boldsymbol{\Delta} - (\bar{\mathbf{x}}_1 - \bar{\mathbf{x}}_2)]^{\mathrm{T}} S^{-1} [\boldsymbol{\Delta} - (\bar{\mathbf{x}}_1 - \bar{\mathbf{x}}_2)] = \frac{p(n_1 + n_2 - 2)}{n_1 + n_2 - p - 1} F_\alpha(p, n_1 + n_2 - p - 1) \tag{7.28}$$

is the equation of a hyperellipsoid centred at $\boldsymbol{\Delta} = \bar{\mathbf{x}}_1 - \bar{\mathbf{x}}_2$ and forms the boundary of a $100(1 - \alpha)$ per cent confidence region for $\boldsymbol{\Delta}$. Again we note that to check if a particular $\boldsymbol{\Delta}^*$ lies within the boundary, we have in effect to carry out a T^2-test on the null hypothesis $H_0: \boldsymbol{\mu}_1 - \boldsymbol{\mu}_2 = \boldsymbol{\Delta}^*$, as in Section 7.5. We see therefore that this approach is not more informative than the test of hypothesis except in the case where $p = 2$, when we can plot the boundary (see Example 7.4 below).

As in Section 7.3, an alternative approach is to find the set of simultaneous confidence intervals such that

$$\mathbf{a}^{\mathrm{T}}(\boldsymbol{\mu}_1 - \boldsymbol{\mu}_2) \in \mathbf{a}^{\mathrm{T}}(\bar{\mathbf{x}}_1 - \bar{\mathbf{x}}_2) \pm K_{\alpha/2} s_a \left[\frac{1}{n_1} + \frac{1}{n_2}\right]^{1/2} \tag{7.29}$$

for all $\mathbf{a}$, with confidence coefficient $100(1-\alpha)$ per cent,

where $s_a^2 = \widehat{\text{Var}}[\mathbf{a}^T\mathbf{X}] = \mathbf{a}^T S \mathbf{a}$ on $(n_1 + n_2 - 2)$ degrees of freedom. (7.30)

and
$$K_{\alpha/2} = \left[\frac{p(n_1+n_2-2)}{n_1+n_2-p-1} F_\alpha(p, n_1+n_2-p-1)\right]^{1/2} \tag{7.31}$$

This again is equivalent to the overall test. When the null hypothesis that $\boldsymbol{\mu}_1 = \boldsymbol{\mu}_2$ is rejected, there is at least one compounding vector $\mathbf{a}$ for which the corresponding confidence interval will not include zero. Again, however, there is no guarantee that the difference between the means of a particular component variate of the p-dimensional random variable will be found significant.

Replacing $K_{\alpha/2}$ by $t_{\alpha/2}(n_1 + n_2 - 2)$ is likely to result in finding more significant discrepancies, but these are more likely to include spurious results.

The linear compound $\mathbf{a}^{*T}\mathbf{X}$ for which the univariate t^2-statistic achieves its maximum value, $\mathscr{T}^2$, should always be examined. As in the one-sample case (leading to Equation (7.4)), it can be shown that $\mathbf{a}^*$ may be taken to be the solution of

$$S\mathbf{a}^* = (\bar{\mathbf{x}}_1 - \bar{\mathbf{x}}_2) \tag{7.32}$$

In a test of hypothesis to compare the two samples of observations on the variable

$$U = \mathbf{a}^{*T}\mathbf{X} \tag{7.33}$$

the corresponding t-statistic is $t_{max} = \mathscr{T}$. This variable was first proposed by R. A. Fisher and is called the *linear discriminant function*. Here we are using it to examine the differences in the means of two populations. Fisher developed it to deal with the problem of allocating an unknown individual to one of two populations where information about these populations is given in samples of observations made on a p-dimensional random variable. This problem will be examined in Section 7.8.

Let us now consider an example to illustrate the procedures of Sections 7.5 and 7.6.

Example 7.4. Table 7.2 gives data for approximately two-year-old girls corresponding to that of Table 7.1 for boys.
Suppose we wish to test the hypothesis that there is no difference in the mean vectors for two-year-old boys and girls.

From Example 7.1 we have

$$n_1 = 6 \qquad \bar{\mathbf{x}}_1^T = [82.0, 60.2, 14.5]$$
$$W_1 = (n_1 - 1)S_1 = \begin{bmatrix} 158.00 & 40.20 & 2.50 \\ 40.20 & 15.86 & 6.55 \\ 2.50 & 6.55 & 9.50 \end{bmatrix}$$

Table 7.2

Individual	Height (cm)	Chest circumference (cm)	MUAC (cm)
1	80	58.4	14.0
2	75	59.2	15.0
3	78	60.3	15.0
4	75	57.4	13.0
5	79	59.5	14.0
6	78	58.1	14.5
7	75	58.0	12.5
8	64	55.5	11.0
9	80	59.2	12.5

From Table 7.2 we calculate

$$n_2 = 9 \qquad \bar{\mathbf{x}}_2^{\mathrm{T}} = [76.0,\ 58.4\ 13.5]$$

$$W_2 = (n_2 - 1)S_2 = \begin{bmatrix} 196.00 & 45.10 & 34.50 \\ 45.10 & 15.76 & 11.65 \\ 34.50 & 11.65 & 14.50 \end{bmatrix}$$

and hence

$$(\bar{\mathbf{x}}_1 - \bar{\mathbf{x}}_2)^{\mathrm{T}} = [6.0,\ 1.8,\ 1.0]$$

$$S = \begin{bmatrix} 27.2308 & 6.5615 & 2.8462 \\ 6.5615 & 2.4323 & 1.4000 \\ 2.8462 & 1.4000 & 1.8462 \end{bmatrix} \text{ on 13 degrees of freedom.}$$

Solving $S\mathbf{a}^* = (\bar{\mathbf{x}}_1 - \bar{\mathbf{x}}_2)$ we obtain

$$\mathbf{a}^{*\mathrm{T}} = [.1278,\ .3493,\ .0797]$$

giving $(\bar{\mathbf{x}}_1 - \bar{\mathbf{x}}_2)^{\mathrm{T}}\mathbf{a}^* = 1.4755$

Hence $\mathscr{T}^2 = 5.3117$

and $$\mathscr{F} = \frac{f - p + 1}{pf}\mathscr{T}^2 = \frac{11}{3.13}\mathscr{T}^2 = 1.50$$

Since $F_{.05}(3, 11) = 3.59$, the difference in means is not significant and we cannot reject the hypothesis that $\boldsymbol{\mu}_1 = \boldsymbol{\mu}_2$ $(P = 0.27)$ □

The interesting point about this example is that if the difference in mean height or in mean chest circumference were to be tested in isolation by the t-test, each test would give a result which is just significant at the 5 per cent level. (Height: $t = 2.18$; chest: 2.19; $t_{.025}(13) = 2.16$) and, of course, $t_{\max} = \mathscr{T} = 2.30$.) Yet, in the multivariate null hypothesis the test statistic was far from significant. This is contrary to the situation discussed in Example 7.2. What is happening here is that the multivariate test is examining simultaneously the set of all possible linear compounds of the three-dimensional variate, and in order to give the required protection against the specified type 1 error, some of the sets in the acceptance region will include contrasts which would prove significant in individual tests. If the third component, MUAC, is omitted so that the set of contrasts being tested contain fewer elements, the multivariate test is still not significant. We find that $\mathscr{T}^2 = 5.29$, $\mathscr{F} = 2.44$; $F_{.05}(2, 12) = 3.89$ $(P = 0.13)$.

Since $p = 2$, we are able to plot the boundary of the 95 per cent confidence region given by Equation (7.28).
Writing $\bar{\mathbf{d}} = \bar{\mathbf{x}}_1 - \bar{\mathbf{x}}_2$, the equation of the boundary is

$$\frac{6 \times 9}{15}(\mathbf{\Delta} - \bar{\mathbf{d}})^{\mathrm{T}} S^{-1}(\mathbf{\Delta} - \bar{\mathbf{d}}) = \frac{2 \times 13}{12} F_{.05}(2, 12)$$

where $S^{-1} = \begin{bmatrix} 0.1049 & -0.2831 \\ -0.2831 & 1.1748 \end{bmatrix}$ and $F_{.05}(2, 12) = 3.8853$

This gives

$$0.1049(\Delta_1 - \bar{d}_1)^2 - 0.5661(\Delta_1 - \bar{d}_1)(\Delta_2 - \bar{d}_2) + 1.1748(\Delta_2 - \bar{d}_2)^2 = 2.3384$$

The boundary is plotted in Fig. 7.1.

The null hypothesis tested above is $H_0 : \mathbf{\Delta} = \mathbf{0}$ (the origin). The point $\mathbf{\Delta} = \mathbf{0}$ lies within the confidence region illustrating the non-significance of the test, the dotted lines denote the limits of the individual confidence intervals for Δ_1 and Δ_2 calculated using the t-distribution on 13 degrees of freedom. The origin $(\mathbf{\Delta} = \mathbf{0})$ lies outside both intervals.

Taken in isolation, the height data supply evidence against the null hypothesis that $\Delta_1 = 0$. The chest data supply evidence against the hypothesis that $\Delta_2 = 0$. Taken together, one piece of evidence does not reinforce the other but, in an apparent paradox, the joint evidence is not sufficient to reject the multivariate null hypothesis. This can be explained to some extent by observing that, taken, together, the two data sets allow an examination of the bivariate structure and it then becomes clear that the observed value of $\bar{\mathbf{d}}$ is not so unlikely given $\mathbf{\Delta} = \mathbf{0}$.

Note that $\mathscr{T}^2$ is always decreased by the removal of a component from the response, but that, if the reduction is small, the value of $\mathscr{F}$ will increase due to

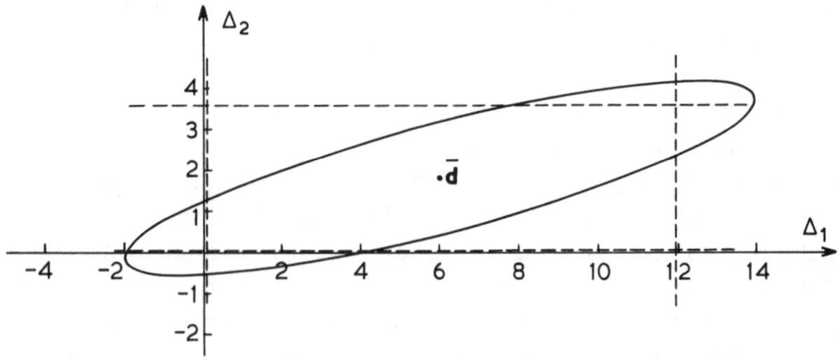

Figure 7.1. Ninety-five per cent confidence region for Δ_1 and Δ_2.

the increase in the factor $(f-p+1)/pf$, with $f = n_1 + n_2 - 2$ in this application.

Here

$$\mathscr{T}_3^2 = 5.31 \qquad \mathscr{F}_3 = \frac{11}{39} \times \mathscr{T}_3^2 = 1.50 \qquad (P = 0.27)$$

$$\mathscr{T}_2^2 = 5.29 \qquad \mathscr{F}_2 = \frac{12}{26} \times \mathscr{T}_2^2 = 2.44 \qquad (P = 0.13)$$

This suggests that a more powerful test is obtained by removing from the analysis any component which supplies no information about the hypothesis being tested. We shall return to this point in Chapter 9.

7.7 Tests of structural relations among the components of the means

In Section 7.4 a test of the hypothesis $H_0: C^{\mathrm{T}}\boldsymbol{\mu} = \boldsymbol{\phi}$ was presented for data arising from a sample from a single population. The approach may be extended to independent samples from two populations having the same covariance matrix Σ, to test the hypothesis $H_0: C^{\mathrm{T}}(\boldsymbol{\mu}_1 - \boldsymbol{\mu}_2) = \boldsymbol{\phi}$. The test statistic is

$$\mathscr{T}^2 = \frac{n_1 n_2}{n_1 + n_2}[C^{\mathrm{T}}(\bar{\mathbf{X}}_1 - \bar{\mathbf{X}}_2) - \boldsymbol{\phi}]^{\mathrm{T}}(C^{\mathrm{T}}SC)^{-1}[C^{\mathrm{T}}(\bar{\mathbf{X}}_1 - \bar{\mathbf{X}}_2) - \boldsymbol{\phi}] \quad (7.34)$$

where S is the pooled estimate of Σ (see Equation (7.24)) and C is a fully specified matrix of order $(p \times m)$ and of rank $m < p$.

Under H_0,

$$\mathscr{T}^2 \sim T_{m}^2(n_1 + n_2 - 2) \quad (7.35)$$

Under the alternative hypothesis,

$$\mathscr{T}^2 \sim T_{m}^2(n_1 + n_2 - 2; \delta^2)$$

where

$$\delta^2 = \frac{n_1 n_2}{n_1 + n_2}[C^{\mathrm{T}}(\boldsymbol{\mu}_1 - \boldsymbol{\mu}_2) - \boldsymbol{\phi}]^{\mathrm{T}}(C^{\mathrm{T}}\Sigma C)^{-1}[C^{\mathrm{T}}(\boldsymbol{\mu}_1 - \boldsymbol{\mu}_2) - \boldsymbol{\phi}] \quad (7.36)$$

Example 7.5. Five male and five female patients suffering from a form of asthma were given a treatment to relieve the condition. The severity of the condition was measured by forced expiratory volume (FEV) and in this trial the FEV was measured at the commencement of treatment and at fixed times during the following two hours. Table 7.3 reports the *increase* in FEV after commencement for each patient, a positive increase denoting an improvement in condition.

The data are to be used to compare the response of males and females to the treatment, and in particular to test the hypothesis that the pattern of the response is sex-dependent. It had been suggested that the effect of the treatment was longer-lasting in females. A cursory examination would

Table 7.3 Increase in FEV ($\times 10^{-2}$ litres)

Sex	Patient	Time after commencement (minutes)				
		5	10	30	60	120
M	1	11	18	15	18	15
	2	33	27	31	21	17
	3	20	28	27	23	19
	4	18	26	18	18	9
	5	22	23	22	16	10
	Mean	20.8	24.4	22.6	19.2	14.0
F	1	18	17	20	18	18
	2	31	24	31	26	20
	3	14	16	17	20	17
	4	25	24	31	26	18
	5	36	28	24	26	29
	Mean	24.8	21.8	24.6	23.2	20.4

support the theory: the improvement in males seems to be falling off at the end of two hours whereas it is being maintained in the female patients. However, there is a large patient variability in the data and we need to test if this difference in pattern is significant.

A univariate analysis is not appropriate since the observations on any one patient may well be correlated.* We shall therefore regard each row of Table 7.3 as an observation on a five-dimensional vector response, so that we have two samples of five observations from the distributions of the response vector generated by the populations of male and female patients. We calculate $\bar{\mathbf{d}}^{\mathrm{T}} = (\bar{\mathbf{x}}_{\mathrm{F}} - \bar{\mathbf{x}}_{\mathrm{M}}) = [4.0, \ -2.6, 2.0, 4.0, 6.4]$ and, using Equation (7.24)

$$S = \begin{bmatrix} 72.700 & 33.025 & 41.650 & 18.675 & 22.300 \\ & 21.250 & 21.300 & 12.725 & 11.925 \\ & & 41.300 & 16.350 & 9.850 \\ & & & 11.450 & 10.200 \\ & & & & 21.650 \end{bmatrix}$$

on 8 degrees of freedom. The lower triangle has been omitted in the covariance matrices throughout this example. Figure 7.2 is a plot of the response for different times.

To investigate the difference in pattern we first ask whether we can reject the null hypothesis that the sex difference is the same at all times. In terms of Fig. 7.2, we are testing to see if the two plots can be considered as essentially parallel. This is, of course, a test of the sex $\times$ times interaction.

* For any one patient the data represents a time series obtained by sampling a continuous process at the given times.

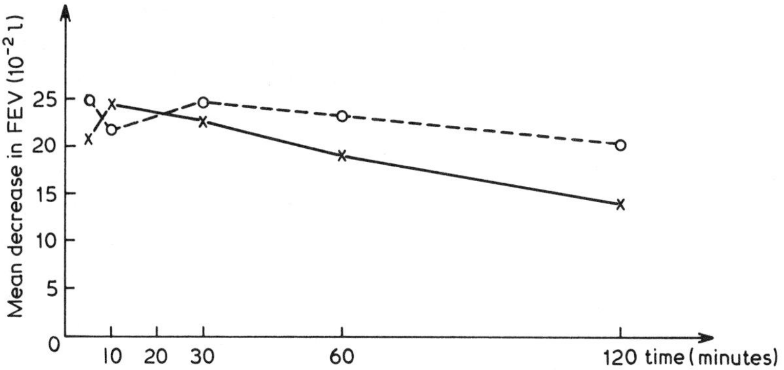

Figure 7.2. Variation of mean decrease in FEV with time. The solid line represents male patients; and the dashed line female patients.

The null hypothesis may be formulated as:

$$H_0: C^T(\boldsymbol{\mu}_F - \boldsymbol{\mu}_M) = \mathbf{0}, \text{ where } C^T = \begin{bmatrix} 1 & -1 & 0 & 0 & 0 \\ 0 & 1 & -1 & 0 & 0 \\ 0 & 0 & 1 & -1 & 0 \\ 0 & 0 & 0 & 1 & -1 \end{bmatrix} \quad (7.37)$$

From Equation (7.34), the test statistic is

$$\mathcal{T}^2 = \frac{n_1 n_2}{n_1 + n_2}(C^T\bar{\mathbf{d}})^T(C^T SC)^{-1}C^T\bar{\mathbf{d}}$$

$$(C^T\bar{\mathbf{d}})^T = [6.6, -4.6, -2.0, -2.4]$$

$$C^T SC = \begin{bmatrix} 27.900 & -8.575 & 14.400 & -4.425 \\ & 19.950 & -16.375 & -5.750 \\ & & 20.050 & 5.250 \\ & & & 12.700 \end{bmatrix}$$

$$(C^T SC)^{-1} = \begin{bmatrix} .091779 & -.031026 & -.107633 & .062547 \\ & .166731 & .158888 & -.001660 \\ & & .278204 & -.081195 \\ & & & .133353 \end{bmatrix}$$

$\mathbf{a}^* = (C^T SC)^{-1}(C^T\bar{\mathbf{d}})$, giving $\mathbf{a}^{*T} = [.81361, -1.28552, -1.80280, .26279]$
Hence $(C^T\bar{\mathbf{d}})^T\mathbf{a}^* = 14.2582$, so that

$$\mathcal{T}^2 = \frac{5 \times 5}{(5+5)} \times 14.2582 = 35.6454$$

$$\mathcal{F} = \frac{f - p + 1}{pf}\mathcal{T}^2_4(8) = \frac{5}{32}\mathcal{T}^2 = 5.57$$

$F_{.05}(4, 5) = 5.19$, and the result is significant at the 5 per cent level ($P = 0.044$). If we write $\mathbf{U} = C^T(\mathbf{X}_1 - \mathbf{X}_2)$, then this analysis is concerned with the four-dimensional random variable **U**. We have tested the hypothesis that $E(\mathbf{U}) = \mathbf{0}$ and rejected it at the 5 per cent level, and in doing so we have found that the linear compound $\mathbf{a}^{*T}\mathbf{U}$ shows the greatest discrepancy between sample and hypothesis.

From the elements of $\mathbf{a}^*$ we can see that the fourth component of **U** contributes least to the value of $\mathcal{T}^2$. This is an indication that the gradients in Fig. 7.2 between 60 and 120 minutes are not of great importance in discriminating between the sexes. In other words, the difference in response between the sexes at 120 minutes is not all that large given the difference that already existed at 60 minutes. The t-statistic to test the hypothesis that these gradients are equal is easily calculated (using the values in $C^T\bar{\mathbf{d}}$ and $C^T SC$) as

$$t^2 = \frac{25}{10} \times \frac{(-2.4)^2}{12.700} = 1.13, \text{ giving } t = 1.06 \text{ (on 8 degrees of freedom)}$$

which would be non-significant in a univariate t-test. Note that the factor 25/10 is

$$\frac{n_1 n_2}{n_1 + n_2} = \left[\frac{1}{n_1} + \frac{1}{n_2}\right]^{-1}$$

as in the expression for $\mathcal{T}^2$.

Further examination of $\mathbf{a}^*$ suggests that the second and third components of **U** (combined using the same sign) are the important ones. We can check this simply by calculating the t-statistic for testing the null hypothesis that $E(\mathbf{a}^T\mathbf{U}) = E(U_2 + U_3) = 0$, where $\mathbf{a}^T = [0, 1, 1, 0]$

$E(\mathbf{a}^T\mathbf{U})$ is estimated by $\bar{u}_2 + \bar{u}_3 = -4.6 - 2.0 = -6.6$ and $\hat{\text{Var}}(\mathbf{a}^T\mathbf{U})$ is estimated by $\mathbf{a}^T \hat{\text{Var}}(\mathbf{U})\mathbf{a}$, where $\hat{\text{Var}}(\mathbf{U}) = C^T SC$, and may be obtained from $C^T SC$ as $19.950 + 20.050 + 2(-16.375) = 7.25$ on 13 degrees of freedom. Hence, the statistic for testing the null hypothesis is

$$t = \frac{-6.6}{\{7.25[(1/5) + (1/5)]\}^{1/2}} = -3.88; \; t_{.005}(13) = 3.01$$

The result is formally significant at the 1 per cent level. It is not significant by the more rigorous K-test (see Equations (7.29)–(7.31); here $K_{.025} = 5.76$).

Having decided that the plots in Fig. 7.2 are not parallel, we could go on to consider if either plot could be regarded as 'flat'. For this we would use the methodology for the first example discussed in Section 7.4 and the test statistic given in Equation (7.19). However, S in Equation (7.19) would refer to the pooled estimate calculated above on 8 degrees of freedom unless we suspected that the assumption of equal covariance matrices for males and females did not hold.

The situation would be different if we had accepted the hypothesis of parallel plots. It would then be sensible to test if the plots were not only parallel but also coincident. The appropriate test statistic would be the t-statistic for testing $H_0: E(\mathbf{a}^T\mathbf{U}) = 0$, where $\mathbf{a}^T = [1, 1, \ldots, 1]$, which tests the difference in the means of the sums of the observations on the components of **X** for the two populations. If a test of 'flatness' of plot seems appropriate, then with parallelism accepted, both plots can be tested together and Equation (7.19) is applied using the grand mean vector (with $n = n_1 + n_2$). □

The series of tests outlined above is called *profile analysis* and is discussed in detail by Morrison (1976), who extends it to consider more than two groups.

It should be stressed that profile analysis should be applied only in those situations where the hypotheses to be tested are relevant *a priori*. Furthermore, since the parallelism of the plots is obviously scale-dependent, the procedure should be restricted to cases where the component variates are measured in the same units. The technique is used to examine profiles (plots) of groups taking a batch of psychological tests, where all the scores are measured on the same scale. The procedure may also be applied where the components consist of some variable measured under different conditions or at different times (as in the example). The scaling problem becomes irrelevant in these circumstances.

7.8 Discriminant analysis

The discrimination problem in its most basic form arises when it is required to allocate an individual to one or other of two populations on the basis of a measurement of a p-dimensional random variable on the individual. It is presumed that the random variable has a different distribution for each of the populations. Denote the density function of the ith population by $f_i(\mathbf{x})$, $i = 1, 2$.

When the losses due to misclassification can be quantified, the problem may be tackled by decision theory. A full treatment may be found in Anderson (1958), who shows that the decision rule is:

$$\left.\begin{array}{l} \text{If } \dfrac{f_1(\mathbf{x})}{f_2(\mathbf{x})} \geq \dfrac{\pi_2 C(1|2)}{\pi_1 C(2|1)}, \text{ allocate the individual giving measurement} \\ \mathbf{x} \text{ to population 1; otherwise allocate to population 2.} \end{array}\right\} \quad (7.38)$$

Here π_i is the prior probability that the individual belongs to population i and $C(i|j)$ is the cost of misallocating an individual from population j to population i.

When $\pi_2 C(1|2) = \pi_1 C(2|1)$, the decision rule is simply to allocate the individual to the population for which the likelihood is greater. When this condition does not apply, the decision rule involves a critical likelihood ratio which depends on the ratio of the misallocation costs and the ratio of prior probabilities.

As an example, consider the use of the decision rule in a screening test for the occurrence of a particular disease, based on observations on a vector random variable. Suppose it is known that 2 per cent of people have the disease and that it is considered that not detecting an individual with the disease is ten times more serious than (provisionally) classifying a healthy individual as diseased. Then if $f_1(\mathbf{x})$ is the density function of the vector random variable in the 'healthy' population and $f_2(\mathbf{x})$ the density function in the 'diseased' population, the decision rule is:

$$\left.\begin{array}{l}\text{If } \dfrac{f_1(\mathbf{x})}{f_2(\mathbf{x})} \geq \dfrac{0.02}{0.98} \times \dfrac{10}{1} \simeq 0.2, \text{ classify as 'healthy';} \\ \text{otherwise recall for further tests.}\end{array}\right\}$$

Note that this rule would imply that a person is not recalled for further tests unless the likelihood ratio is *more* than five to one in favour of allocation to the 'diseased' population. If this consequence were not to prove acceptable to the doctors running the tests, the likely implication would be that the misclassification costs had been wrongly estimated.

In the case of normal distributions, it is possible to derive a simplified verson of the rule. Suppose, for the ith population,

$$\mathbf{X} \sim \mathrm{N}_p(\boldsymbol{\mu}_i, \Sigma) \qquad i = 1, 2$$

where we note the assumption of equal covariances. Then

$$f_i(\mathbf{x}) = (2\pi)^{-p/2}|\Sigma|^{-1/2}\exp\{-\tfrac{1}{2}(\mathbf{x}-\boldsymbol{\mu}_i)^{\mathrm{T}}\Sigma^{-1}(\mathbf{x}-\boldsymbol{\mu}_i)\}$$

and so, after some algebra,

$$\ln\left[\frac{f_1(\mathbf{x})}{f_2(\mathbf{x})}\right] = \mathbf{x}^{\mathrm{T}}\Sigma^{-1}(\boldsymbol{\mu}_1-\boldsymbol{\mu}_2) - \tfrac{1}{2}(\boldsymbol{\mu}_1-\boldsymbol{\mu}_2)^{\mathrm{T}}\Sigma^{-1}(\boldsymbol{\mu}_1+\boldsymbol{\mu}_2) \tag{7.39}$$

Writing $\mathbf{L} = \Sigma^{-1}(\boldsymbol{\mu}_1 - \boldsymbol{\mu}_2)$ and $k = \ln[(\pi_2 C(1|2))/(\pi_1 C(2|1))]$, we have the decision rule:

$$\left.\begin{array}{l}\text{If } \mathbf{L}^{\mathrm{T}}\mathbf{x} - \dfrac{1}{2}\mathbf{L}^{\mathrm{T}}(\boldsymbol{\mu}_1+\boldsymbol{\mu}_2) \geq k, \text{ allocate to population 1;} \\ \text{otherwise allocate to population 2.}\end{array}\right\} \tag{7.40}$$

This decision rule is based on Fisher's linear discriminant function which was mentioned earlier in Section 7.6 (see Equation (7.33) and also Equation (7.52) below). Note that $k = 0$ when $\pi_2 C(1|2) = \pi_1 C(2|1)$.

Now define
$$U = \mathbf{L}^{\mathrm{T}}\mathbf{X} - \tfrac{1}{2}\mathbf{L}^{\mathrm{T}}(\boldsymbol{\mu}_1+\boldsymbol{\mu}_2) \tag{7.41}$$

and assume $\mathbf{L}^{\mathrm{T}}\boldsymbol{\mu}_1 > \mathbf{L}^{\mathrm{T}}\boldsymbol{\mu}_2$

Then if $\mathbf{X}$ comes from population 1 we have

$$U = U_1 \sim \mathrm{N}_1(\tfrac{1}{2}\alpha, \alpha) \tag{7.42}$$

whereas if $\mathbf{X}$ comes from population 2, we have

$$U = U_2 \sim \mathrm{N}_1(-\tfrac{1}{2}\alpha, \alpha) \tag{7.43}$$

where $\alpha = (\boldsymbol{\mu}_1-\boldsymbol{\mu}_2)^{\mathrm{T}}\Sigma^{-1}(\boldsymbol{\mu}_1-\boldsymbol{\mu}_2) = \mathbf{L}^{\mathrm{T}}(\boldsymbol{\mu}_1-\boldsymbol{\mu}_2)$ (7.44)

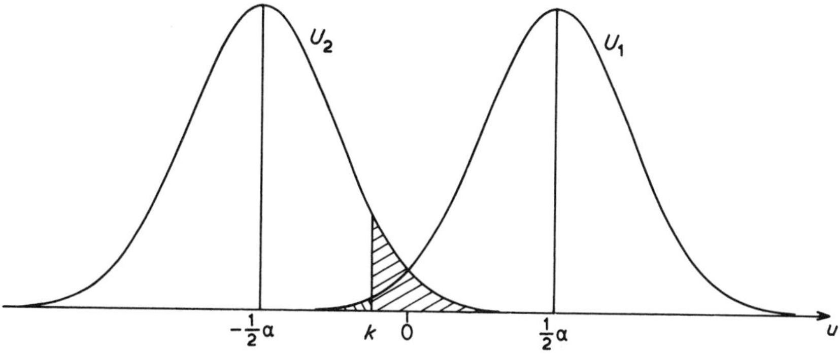

Figure 7.3. Distributions of U from the two populations. Shaded areas represent the conditional probabilities of misclassification, $P(i|j)$, for decision point, k.

α is a measure of distance between the two population means, and is known as the *Mahalanobis distance.*

The situation may be represented pictorially as in Fig. 7.3.

The shaded areas in Fig. 7.3 represent the conditional probabilities of misclassification, $P(i|j)$, where $P(i|j)$ = Probability [individual from population j is misallocated to population i]

Then

$$\left.\begin{aligned} P(1|2) &= P(U_2 > k) \\ P(2|1) &= P(U_1 < k) \end{aligned}\right\} \tag{7.45}$$

Note that these conditional probabilities are equal when $k = 0$

The *total* probability of misclassification is

$$\pi_1 P(2|1) + \pi_2 P(1|2) \tag{7.46}$$

When $k = 0$, the total probability of misclassification is

$$\begin{aligned} &\pi_1 P(U_1 < 0) + \pi_2 P(U_2 > 0) \\ &= P(U_1 < 0) \\ &= P\left[\frac{U_1 - \frac{1}{2}\alpha}{\sqrt{\alpha}} < \frac{-\frac{1}{2}\alpha}{\sqrt{\alpha}}\right] \\ &= \Phi(-\tfrac{1}{2}\sqrt{\alpha}) \end{aligned} \tag{7.47}$$

where $\Phi(x)$ is the distribution function of the standard normal distribution.

It is interesting to note that the decision point can be easily represented pictorially. Suppose Fig. 7.3 is redrawn so that each of the two distributions is scaled by multiplying the ordinates of the density function of U_i $(i = 1, 2)$ by $\pi_i C(j|i)$ as is shown in Fig. 7.4 for a particular scaling.

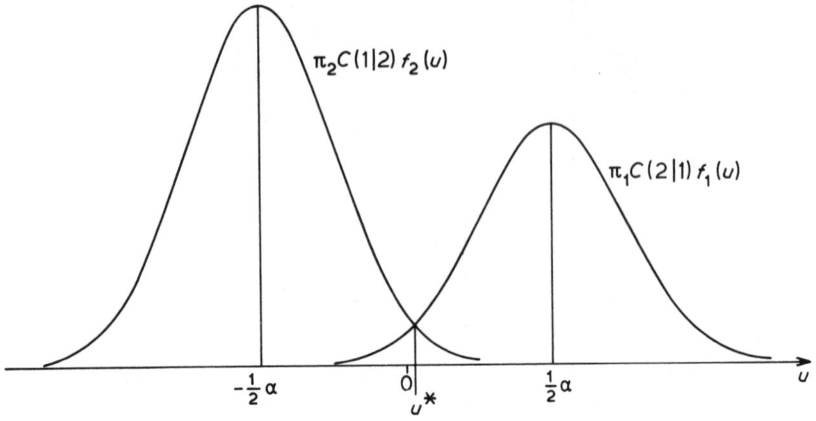

Figure 7.4. Plots of $\Pi_i\, C(j\,|\,i) f_i(u)$ for $i,j = 1,2$.

At $u = u^*$, the point where the density functions intersect,

$$\pi_2 C(1|2) f_2(u^*) = \pi_1 C(2|1) f_1(u^*) \tag{7.48}$$

where $f_i(u) = (1/\sqrt{2\pi\alpha})\exp[-(u \pm \frac{1}{2}\alpha)^2/2\alpha]$ from Equations (7.42) and (7.43). Substituting in Equation (7.48) and taking natural logarithms, this reduces to $\ln(\pi_2 C(1|2)) - (u^* + \frac{1}{2}\alpha)^2/2\alpha = \ln(\pi_1 C(2|1)) - (u^* - \frac{1}{2}\alpha)^2/2\alpha$ and after some simplification to

$$u^* = \ln\left[\frac{\pi_2 C(1|2)}{\pi_1 C(2|1)}\right] = k \tag{7.49}$$

When the prior probabilities are unknown, Anderson (1958) shows that a minimax decision rule is applicable and that it is determined by finding k, the decision point, such that the costs of the two types of misclassification are equal. That is,

$$C(2|1) P(U_1 < k) = C(1|2) P(U_2 > k) \tag{7.50}$$

A solution of Equation (7.50) may be obtained by trial and error using tables of the normal distribution. Note that when $C(2|1) = C(1|2)$ the minimax decision point is at $k = 0$.

The discriminant function is linear in $\mathbf{x}$ in the case of equal covariance matrices because the quadratic terms cancel when the log-likelihood ratio is calculated. When the covariance matrices are not equal, the discriminant function will be quadratic in $\mathbf{x}$. The decision rule will consequently be more complicated and the problem is then best tackled using *discriminant scores* as suggested by Rao (1973, Section 8e).

Up to this point in the presentation, we have assumed that the distributions of $\mathbf{X}$ over the two populations are known. In practice, the forms of the distributions are assumed and the parameters estimated from data where we know from which population each observation comes (either *a priori* or perhaps by some more expensive or inconvenient method of identification). The decision rule is then taken to be Equation (7.38), or (7.40) in the case of two normal populations (with equal covariance matrices) but with the parameters replaced by their estimates. For example, Equation (7.40) would be replaced in practice by:

$$\left.\begin{array}{l}\text{If } \mathbf{L}^{\mathrm{T}}\mathbf{x} - \tfrac{1}{2}\mathbf{L}^{\mathrm{T}}(\bar{\mathbf{x}}_1 + \bar{\mathbf{x}}_2) \geq k, \text{ allocate to population 1;} \\ \text{otherwise allocate to population 2}\end{array}\right\} \qquad (7.51)$$

Here
$$\mathbf{L} = S^{-1}(\bar{\mathbf{x}}_1 - \bar{\mathbf{x}}_2) \qquad (7.52)$$

and S is the pooled within-groups estimate of Σ. On comparing this with Equation (7.33), we see that $\mathbf{L} = \mathbf{a}^*$, the vector which maximizes the t^2-statistic for comparing the two population means, and $\mathbf{L}^{\mathrm{T}}\mathbf{x}$ is Fisher's linear discriminant function.

An immediate consequence of this is that the distributions of U_1 and U_2 defined in Equations (7.42) and (7.43) are now very complicated, as also is the estimation of the probabilities of misclassification. The distributional problems are discussed in Anderson (1958) and in Morrison (1976), who also discusses methods for estimating the probabilities of misclassification.

The evaluation of k also presents problems. If the sample sizes n_i in the data used to construct the discriminant function are thought to reflect the unknown prior probabilities of occurrence of individuals from the two populations, π_i may be estimated by $n_i/(n_1 + n_2)$, $i = 1, 2$, but otherwise the minimax decision point should be estimated, using Equation (7.50). The assessment of misclassification costs may also be difficult so that there is a temptation to measure the efficiency of discrimination solely in terms of the probability (and not the cost) of misclassification. In many applications this may be quite satisfactory, as for example in the problem of allocating an individual plant specimen to one species or another, but in other applications misclassification costs are too important to ignore. The medical screening example is a case in point, and so is the problem of identifying a site as either potentially valuable in some geological sense or not. Sometimes the decision point may be arrived at pragmatically. In the case of the medical screening example, the decision point might be adjusted with experience so that the number of individuals recalled for further tests is just within the capabilities of the available resources.

The pictorial representation in Fig. 7.4 leading to the identification of the decision point as the intersection of the scaled densities in Equation (7.49), may also be used to suggest approximate adjustments to the decision point where histograms of the discriminant values for the original data suggest that the assumption of equal covariances is suspect. Strictly, a quadratic discriminant rule should be estimated but the linear discriminant rule is so simple to use that it may be considered acceptable although sub-optimal.

The decision rule of Equation (7.38) may be recast to allocate the individual to the population for which the value of the discriminant score,

$w_i = -\Sigma \pi_j C(i|j) f_j(\mathbf{x})$ is greater (take $C(i|i) = 0$ for $i = 1, 2$). This approach, which enables the discriminant theory to be extended to more than two populations, is clearly presented in Rao (1973). In the case of normal populations with equal covariances and equal misclassification costs, it can be shown that equivalent discriminant scores are calculated from

$$w_i = \mathbf{L}_i^{\mathrm{T}}\mathbf{x} - \frac{1}{2}\mathbf{L}_i^{\mathrm{T}}\bar{\mathbf{x}}_i + \ln(\pi_i) \qquad i = 1, 2, \ldots, m \tag{7.53}$$

where m is the number of populations, $\bar{\mathbf{x}}_i$ is the mean vector for population i, S is the pooled within-groups estimate of the common Σ and $\mathbf{L}_i = S^{-1}\bar{\mathbf{x}}_i$. Note that in the previous approach with two populations only, $\mathbf{L}$ is related to the $\mathbf{L}_i$ by $\mathbf{L} = \mathbf{L}_1 - \mathbf{L}_2$ (see Equation (7.52)).

The procedure is to calculate Equation (7.53) for each population and to allocate the individual to that population for which w_i is greatest. It forms the basis for most computer programs for discriminant analysis. An alternative method using Mahalanobis distances given by Morrison (1976) is essentially the same but requires slightly heavier calculation. A review of recent work on discriminant analysis is given by Lachenbruch and Goldstein (1979).

In the next chapter, an alternative approach for discriminating among more than two populations will be based on *canonical variates* analysis.

Exercises

7.1. If $\mathbf{u}$ and $\mathbf{v}$ are real vectors, the Cauchy–Schwartz inequality may be stated as

$$(\mathbf{u}^{\mathrm{T}}\mathbf{v})^2 \leq (\mathbf{u}^{\mathrm{T}}\mathbf{u})(\mathbf{v}^{\mathrm{T}}\mathbf{v})$$

with equality if and only if $\mathbf{u} = c\mathbf{v}$ for some real scalar c. Suppose S is a real symmetric matrix of full rank, then we can find a non-singular matrix B such that $S = BB^{\mathrm{T}}$. In the Cauchy–Schwartz inequality let $\mathbf{u} = B^{\mathrm{T}}\mathbf{a}$ and $\mathbf{v} = B^{-1}(\bar{\mathbf{X}} - \boldsymbol{\mu}_0)$. Hence derive Equation (7.4) and give the necessary and sufficient condition for achieving the maximum.

7.2. Using the data in Table 7.2, test the following hypotheses on the population of girls:

(i) $\boldsymbol{\mu}^{\mathrm{T}} = [80, 60, 15]$

(ii) $\frac{1}{5}\mu_1 = \frac{1}{4}\mu_2 = \mu_3$

7.3 (*a*) A p-dimensional variate consists of measurements on a response of an experimental unit at *equal* time intervals. An hypothesis states that the mean response varies linearly with time. This may be stated as $H_0: \mu_i = \alpha + \beta t_i$, $i = 1, 2, \ldots, p$.

Show that under this hypothesis the expected values of the second differences of the response are zero, so that the hypothesis may be restated as $H_0: C^{\mathrm{T}}\boldsymbol{\mu} = \mathbf{0}$, where $C = \begin{bmatrix} 1 & 0 & \dots & 0 \\ -2 & 1 & & \\ 1 & -2 & & \vdots \\ 0 & 1 & & \\ & & & 1 \\ \vdots & & & -2 \\ 0 & & & 1 \end{bmatrix}$

How would you test H_0?

7.3. (*b*) Show how the above procedure could be modified to cope with unequal time intervals. Apply the method to test the hypothesis that for the male patients in Table 7.3, the increase in FEV decreases linearly with time for the given range.

7.4. Carry out a profile analysis to compare boys and girls using the data given in Tables 7.1 and 7.2.

7.5. Let us redefine the response variate and the two populations for the data in Table 7.3. Suppose the response consists of the results of a series of five diagnostic tests and the first sample refers to the population of individuals suffering from a serious disease which is difficult to diagnose. The second sample is from a population with a less serious complaint but having very similar basic symptoms. Does the mean response for the five tests differ for the two groups? Construct Fisher's linear discriminant function to distinguish between the two complaints. Using the decision rule, how many of the ten patients would have been correctly diagnosed? If it is known that about 10 per cent of patients showing the basic symptoms have the more serious disease, how should the decision point be modified?

CHAPTER EIGHT

The multivariate analysis of variance

8.1 Introduction

In this chapter the consideration of procedures based on normal distribution theory will be extended to the analysis of data arising from designed experiments. We shall find that the calculation of a multivariate analysis of variance (MANOVA) is essentially similar to the calculation of a univariate analysis of variance (ANOVA). However, the testing of hypotheses and the interpretation of results is more complicated.

8.2 MANOVA calculations

In a designed experiment, observations are made following a plan which specifies the conditions under which each experimental unit is selected for measurement. For a multivariate analysis, p variates are measured on each experimental unit, giving a vector response, and not just the one (scalar) response as is the case for a univariate analysis. Note that the experimental plan or design specifies the selection of experimental units and so will be the same for each component of the response vector.

In a univariate analysis of variance, the total sum of squared deviations about the grand mean – written SS (Total) – is partitioned into a sum of squares due to one or more sources and a residual sum of squares. Associated with each partition is a number, called the *degrees of freedom* (d.f.), representing the number of linearly independent contrasts or alternatively the number of linearly independent parameters for that source. The partitioning is set out in an analysis-of-variance table.

In a p-dimensional multivariate analysis of variance based on the same design there are, p SS (Total)'s to partition, one for each component measured. In addition, there are measures of covariance between pairs of components presented as sums of products. The MANOVA calculation is concerned with the partition of these measures of variance and covariance which are collected in a matrix of sums of squares and products written SSPM (Total). This is partitioned into sums of squares and product matrices due to the same sources as in the univariate case, and a residual sum of squares and products matrix. The matrices are all symmetric.

Table 8.1 Data for Example 8.1

Block	Treatment 1	Treatment 2	Treatment 3	Total
1	$\begin{bmatrix}13.3\\10.6\\21.2\end{bmatrix}$	$\begin{bmatrix}13.6\\10.2\\21.0\end{bmatrix}$	$\begin{bmatrix}14.2\\10.7\\21.1\end{bmatrix}$	$\begin{bmatrix}41.1\\31.5\\63.3\end{bmatrix}$
2	$\begin{bmatrix}13.4\\9.4\\21.0\end{bmatrix}$	$\begin{bmatrix}13.2\\9.6\\20.1\end{bmatrix}$	$\begin{bmatrix}13.9\\10.4\\19.8\end{bmatrix}$	$\begin{bmatrix}40.5\\29.4\\60.9\end{bmatrix}$
3	$\begin{bmatrix}12.9\\10.0\\20.5\end{bmatrix}$	$\begin{bmatrix}12.2\\9.9\\20.7\end{bmatrix}$	$\begin{bmatrix}13.9\\11.0\\19.1\end{bmatrix}$	$\begin{bmatrix}39.0\\30.9\\60.3\end{bmatrix}$
Total	$\begin{bmatrix}39.6\\30.0\\62.7\end{bmatrix}$	$\begin{bmatrix}39.0\\29.7\\61.8\end{bmatrix}$	$\begin{bmatrix}42.0\\32.1\\60.0\end{bmatrix}$	$\begin{bmatrix}120.6\\91.8\\184.5\end{bmatrix}$

Since the design of the experiment is the same, the degrees of freedom will be the same in the MANOVA as in the ANOVA.

The basic (non-computer) method of calculation depends on the fact that the numbers in a particular position on the diagonals of the matrices are simply the sums of squares calculated for an ANOVA of the corresponding component of the vector response. For the off-diagonal terms, essentially the same formulae are used but squared terms are replaced by product terms for each pair of components. An example should make this clear.

Example 8.1. Suppose three treatments are compared in a randomized block experiment with three blocks as in Table 8.1. The response is trivariate. For component 1:

$$\text{SS (Total)} = 13.3^2 + 13.6^2 + \ldots + 12.2^2 + 13.9^2 - 120.6^2/9 = 2.92$$

$$\text{SS (Treatments)} = (39.6^2 + 39.0^2 + 42.0^2)/3 - 120.6^2/9 = 1.68$$

$$\text{SS (Blocks)} = (41.1^2 + 40.5^2 + 39.0^2)/3 - 120.6^2/9 = 0.78$$

Hence SS (Residual) = 0.46, by subtraction.

Covariance terms for components 1 and 2 are calculated using analogous expressions

$$\text{SP (Total)} = (13.3 \times 10.6) + (13.6 \times 10.2) + \ldots + (12.2 \times 9.9) + (13.9 \times 11.0) - (120.6 \times 91.8)/9 = 1.44$$

$$\text{SP (Treatments)} = \{(39.6 \times 30.0) + (39.0 \times 29.7) + (42.0 \times 32.1)\}/3 - (120.6 \times 91.8)/9 = 1.38$$

$$\text{SP (Blocks)} = \{(41.1 \times 31.5) + (40.5 \times 29.4) + (39.0 \times 30.9)\}/3 - (120.6 \times 91.8)/9 = 0.03$$

Hence SP(Residual) = 0.03

Here the expressions for sums of squares are calculated in terms of totals. To obtain sums of products, each squared total in a sum of squares is replaced by the product of the totals corresponding to the two components involved. Otherwise the expressions are the same. However, remember that *sums of products can be negative.*

Complete the calculations and check your results against the MANOVA table (Table 8.2). □

Table 8.2 MANOVA table for Example 8.1

Source	d.f.	SSPM
Treatments	2	$\begin{bmatrix} 1.68 & 1.38 & -1.26 \\ 1.38 & 1.14 & -1.08 \\ -1.26 & -1.08 & 1.26 \end{bmatrix}$
Blocks	2	$\begin{bmatrix} 0.78 & 0.03 & 0.96 \\ 0.03 & 0.78 & 0.66 \\ 0.96 & 0.66 & 1.68 \end{bmatrix}$
Residual	4	$\begin{bmatrix} 0.46 & 0.03 & -0.40 \\ 0.03 & 0.30 & -0.48 \\ -0.40 & -0.48 & 1.06 \end{bmatrix}$
Total	8	$\begin{bmatrix} 2.92 & 1.44 & -0.70 \\ 1.44 & 2.22 & -0.90 \\ -0.70 & -0.90 & 4.00 \end{bmatrix}$

This similarity between MANOVA and ANOVA holds for any design. The amount of calculation is, of course, much greater. In this example with a three-dimensional response, the calculation of the MANOVA is the equivalent of six ANOVA calculations. For a p-dimensional response, the calculation would be the equivalent of $\frac{1}{2}p(p+1)$ ANOVA calculations. For a large experiment with a high dimensional response, the calculations will be so heavy that a computer is practically essential.

It should be noted that on a computer the calculations are carried out by matrix operations on the data matrix. Since standard computer packages are universally available, it is not proposed to give the matrix formulation of the SSPM. Even for the simpler designs these involve a complicated notation which detracts from the basic simplicity of the methodology.

Before proceeding to the hypothesis-testing stage, here are two further data sets to provide practice in the calculations. The examples are highly simplified versions of real experiments.

Example 8.2. Prior to experimenting with treatments to inhibit tumour growth, a fully randomized experiment was carried out to investigate the effect of environmental temperature on the growth of tumours seeded sub-

Table 8.3 Data for Example 8.2

Temperature (°C)	Male Replicates			Male Mean	Female Replicates			Female Mean	Mean
4	18.15 16.51 0.24	18.68 19.50 0.32	19.54 19.84 0.20	18.790 18.617 0.253	19.15 19.49 0.16	18.35 19.81 0.17	20.68 19.44 0.22	19.393 19.580 0.183	19.092 19.098 0.218
20	21.27 23.30 0.33	19.57 22.30 0.45	20.15 18.95 0.35	20.330 21.517 0.377	18.87 22.00 0.25	20.66 21.08 0.20	21.56 20.34 0.20	20.363 21.140 0.217	20.347 21.328 0.297
34	20.74 16.69 0.31	20.02 19.26 0.41	17.20 15.90 0.28	19.320 17.283 0.333	20.22 19.00 0.18	18.38 17.92 0.30	20.85 19.90 0.17	19.817 18.940 0.217	19.568 18.112 0.275
Mean				19.480 19.139 0.321				19.858 19.887 0.206	19.669 19.513 0.263

cutaneously in laboratory-bred mice. Three mice of each sex were reared in each of three environmental temperatures (see Table 8.3). Variates measured were X_1, the initial weight, X_2, the final weight (less the tumour weight), and X_3, the tumour weight. All weights were in grammes.

Example 8.3. Bivariate data were collected in a fully randomized experiment. The variates observed were X_1, total seed yield per plant in grammes, and X_2, the weight of 100 randomly sampled seeds per plant in milligrammes (see Table 8.4). The four treatments were equally spaced levels of a nitrogenous fertilizer. □

The MANOVA table for Example 8.2 is given in the next section (Table 8.7). Example 8.3 is set as an exercise at the end of the chapter and the answers will be found there.

Table 8.4 Data for Example 8.3

Nitrogen level	Replicates				Treatment total
1	21 12	25 8	20 12	24 10	90 42
2	31 9	23 12	24 13	28 10	106 44
3	34 10	29 14	35 11	32 13	130 48
4	33 14	38 12	34 13	35 13	140 52

8.3 Testing hypotheses

8.3.1 *The special case: The univariate procedure*

Before discussing multivariate procedures, let us review the univariate case by considering the data in Example 8.1 for the first component only.

The full model to be fitted is a linear model containing block and treatment effects. The model, denoted by Ω, may be expressed as

$$\Omega: X_{ij} = \mu + \alpha_i + \beta_j + e_{ij} \qquad i = 1, 2, 3;\ j = 1, 2, 3 \tag{8.1}$$

where the α_i are the treatment effects, β_j are the block effects, and e_{ij} are assumed normally and independently distributed with mean zero and variance σ^2. All contrasts of interest – among treatments and between blocks – are estimable.

The usual null hypothesis to be tested is $H_0 : \alpha_1 = \alpha_2 = \alpha_3$. stating that the response is the same for all treatment levels. Under this hypothesis the model Ω is reduced to model Ω_0, written

$$\Omega_0 : X_{ij} = \mu + \beta_j + e_{ij} \tag{8.2}$$

The calculations for the test are set out in Table 8.5 as an ANOVA table (extracted from Table 8.2).

Under the null hypothesis, the variance ratio is distributed as $F(2, 4)$ and the observed value is compared with the percentage points of that distribution. Here $7.30 > F_{.05}(2, 4) = 6.94$, so that the treatment effect is significant at the 5 per cent level.

The test is an example of the test of a linear hypothesis in the theory of least squares and it is instructive to reformulate it in terms of the general theory. The residual sum of squares in Table 8.5 is the residual after fitting block and treatment effects. Denote it by $R(\Omega)$. Under the null hypothesis that there are no treatment effects, the model Ω_0 requires the fitting of block effects only. Since the total sum of squares does not change and, because of the orthogonality of the design, the 'blocks' sum of squares is also unaltered, the residual sum of squares for the reduced model Ω_0, denoted by $R(\Omega_0)$ is

$$R(\Omega_0) = R(\Omega) + \text{SS (Treatments)} \tag{8.3}$$

with corresponding degrees of freedom

$$6 = 4 + 2 \tag{8.4}$$

Table 8.5

Source	d.f.	Sums of squares	Mean square	Variance ratio
Treatments	2	1.68	0.840	7.30
Blocks	2	0.78		
Residual	4	0.46	0.115	
Total	8	2.92		

The test statistic given by the general theory is

$$\frac{[R(\Omega_0) - R(\Omega)]/(r_0 - r)}{R(\Omega)/r} \tag{8.5}$$

where $r =$ d.f. for R, and $r_0 =$ d.f. for R_0. In this example $r = 4$, $r_0 = 6$. $R(\Omega_0) - R(\Omega)$ is called the sum of squares for testing the hypothesis. The test statistic is invariant under changes of origin and scale and, with the distributional assumptions of the model, the test is known to be uniformly most powerful against the alternative hypothesis that there are treatment effects.

8.3.2 *The multivariate model for Example* 8.1

Now let us consider the multivariate case. The model Ω now becomes

$$\Omega : \mathbf{X}_{ij} = \boldsymbol{\mu} + \boldsymbol{\alpha}_i + \boldsymbol{\beta}_j + \mathbf{e}_{ij} \qquad \mathrm{i} = 1, 2, 3;\ j = 1, 2, 3 \tag{8.6}$$

where the parameters are now (3×1) vectors and the $\mathbf{e}_{ij}$ are vector random variables assumed to be distributed independently as $\mathrm{N}_3(\mathbf{0}, \Sigma)$. Note that the 'errors' for different experimental units will be uncorrelated but that 'errors' from the same experimental unit may be correlated.

Under the null hypothesis $H_0 : \boldsymbol{\alpha}_i = \mathbf{0}$, $i = 1, 2, 3$, the reduced model is

$$\Omega_0 : \mathbf{X}_{ij} = \boldsymbol{\mu} + \boldsymbol{\beta}_j + \mathbf{e}_{ij} \tag{8.7}$$

The MANOVA table (Table 8.2) may be written as in Table 8.6. where H, B, R *and* T denote the (3×3) matrices calculated in Section 8.2. The residual sum of squares and products matrix for the reduced model Ω_0 is denoted by R_0, and

$$R_0 = R + H \qquad \text{on} \quad (r + h)\text{d.f.} \tag{8.8}$$

Multivariate test procedures are based on comparisons of the matrices R and R_0.

One complication of a multivariate analysis that does not arise in the univariate case has to do with the ranks of the matrices. It was shown in Section 6.5 that the rank of any matrix on f d.f. in a p-variate MANOVA is theoretically equal to the lesser of f or p. As an instance of this, it is easily verified in Example 8.1 that rank $(H) = 2$. In practice, because of rounding errors in the calculations the rank will usually be p even when $f < p$, although

Table 8.6

Source	d.f.	SSPM
Treatments	h	H
Blocks	b	B
Residual	r	R
Total	t	T

then $(p - f)$ of the eigenvalues of the matrix will be small. It is even possible for the rank to be less than the theoretical value due to the fact that the data are recorded to a fixed number of decimal places. Suppose two of the treatment mean vectors in Example 8.1 were equal, an event which in the theory is considered to have zero probability, then (with no rounding error) it would be found that rank $(H) = 1$.

In the remaining part of this chapter, the theoretical rank of any hypothesis matrix on h d.f. will be denoted by s, where

$$s = \min(h, p) \tag{8.9}$$

It will nearly always be found that R is of full rank, p, so long as its degrees of freedom $r \geq p$, but see the note following Equation (8.10) and the discussion in Section 3.2.2.

8.3.3 *Multivariate test procedures*

Anderson (1958, Section 8.10) discusses the choice of test procedures and shows that the only statistics invariant under choice of origin and scale of the data are the roots† of the equation

$$|H - \theta R| = 0 \tag{8.10}$$

This is a polynomial in θ of degree p. It can be shown that the roots θ_i, $i = 1, 2, \ldots, p$, satisfy $\theta_i \geq 0$, but if s (= rank H) $< p$ then $p - s$ of the roots are zero. In Example 8.1, $p = 3$, $s = 2$ and the roots are $\theta_1 = 145.613$, $\theta_2 = 0.581535$ with $\theta_3 = 0$.

The θ_i are the eigenvalues of HR^{-1}. This follows by rewriting Equation (8.10) as

$$|HR^{-1} - \theta I| = 0 \tag{8.11}$$

Note that HR^{-1} is not symmetric.

It is not immediately obvious that the statistics given by Equation (8.10) are intuitively sensible. However, when $p = 1$ the single eigenvalue θ_1 is equal to H/R, which is proportional to the variance-ratio statistic which arises in a univariate ANOVA.

A number of test procedures have been based on functions of the θ_i. One possible test statistic is θ_1, the greatest root of Equation (8.10). This was proposed by S. N. Roy and is the statistic used by Morrison (1976). Another

† It should be pointed out that if R is not of full rank, Equation (8.10) holds for all values of θ. This situation will arise if the degrees of freedom for R are less than p, but can easily be avoided by forethought when planning the experiment. It can also arise if at least one linear relationship among the components of the response vector holds for every individual observation (and consequently for the mean vectors giving rise to the matrix H). This second case can be overlooked due to rounding errors in the calculation of the SSPM. It is therefore worthwhile to examine the rank of R by evaluating its eigenvalues. If one or more is very small, little information is lost by omitting a corresponding number of the components of the response.

statistic is

$$T = \text{trace}\,(HR^{-1}) = \sum_{i=1}^{s} \theta_i \qquad (8.12)$$

proposed by D. Lawley and H. Hotelling.

The test statistic to be used in this book is the Wilks' Λ-statistic developed by S. S. Wilks using the likelihood-ratio approach, namely

$$\Lambda = \frac{|R|}{|H + R|} \qquad (8.13)$$

It is easy to show that Λ can be expressed as a function of the θ_i. Rewrite Equation (8.10) as

$$|(1 + \theta)R - (H + R)| = 0$$

giving

$$|R(H + R)^{-1} - \lambda I| = 0 \qquad (8.14)$$

where $\lambda = 1/(1 + \theta)$
Then, using Equation (1.14),

$$\Lambda = \frac{|R|}{|H + R|} = |R(H + R)^{-1}| = \prod_{i=1}^{p} \lambda_i = \prod_{i=1}^{p} \frac{1}{(1 + \theta_i)} \qquad (8.15)$$

If $s = \text{rank}\,(\text{H}) < p$, then $\theta_{s+1} = \theta_{s+2} = \ldots = \theta_p = 0$ and correspondingly $\lambda_{s+1} = \lambda_{s+2} = \ldots = \lambda_p = 1$.

Note that the λ_i, the eigenvalues of $R(H + R)^{-1}$, are such that $0 \leq \lambda_i \leq 1$. Small values of λ (corresponding to large values of θ) provide evidence against the null hypothesis.

We can check Equation (8.15) using Example 8.1

$|R| = 0.002862$ and $|H + R| = 0.663624$, giving $\Lambda = 0.00431268$ and

$$\prod_{i=1}^{s} \frac{1}{1 + \theta_i} = (146.613 \times 1.581535)^{-1} = 0.00431269$$

Two other statistics are

$$V = \text{trace}\,[H(H + R)^{-1}] = \sum_{i=1}^{s} \frac{\theta_i}{1 + \theta_i} \qquad (8.16)$$

proposed by K. C. S. Pillai, and

$$U = \prod_{i=1}^{s} \frac{\theta_i}{1 + \theta_i} \qquad (8.17)$$

proposed by Roy *et al.* (1971). This is the product of the non-zero eigenvalues of $H(H + R)^{-1}$

Power comparisons among these and other statistics have been made by Schatzoff (1966), Pillai and Jayachandran (1967), Roy *et al.* (1971) and Lee

(1971). The results of these and other studies are indecisive. Which statistic is more powerful depends on the structure of the alternative hypothesis. In particular, θ_1 is best when the alternative specifies that the treatment means are collinear or nearly so. Λ, T and V are asymptotically equivalent and differ little in power for small samples. Olsen (1974) has compared the statistics for their properties of robustness and he slightly favours V, although again there is little difference. He is against the use of Λ in MANOVA when the equality of covariance structures does not hold for the different treatments. However, Λ has been chosen for this book partly because of its ease of calculation in problems of low dimensionality, but mainly for the existence of distributional approximations which enable (approximate) critical values to be readily found. The tests are all exactly equivalent when $p = 1$.

8.3.4 *Distributional approximations*

Assuming the normal distribution, Rao (1973) shows that under the null hypothesis, Λ is distributed as the product of independent beta variables, but for practical purposes an approximation† due to Bartlett (1947) is more useful. With the notation of Table 8.6, write

$$\Lambda_{p,h,r} = \frac{|R|}{|H+R|}$$

then

$$-\left(r - \frac{p-h+1}{2}\right)\ln\Lambda \sim \chi^2(ph) \tag{8.18}$$

where the premultiplier of $\ln\Lambda$ is chosen to adjust the asymptotic distribution for finite r in an optimum way (see Anderson (1958), Section 8.6).

Exact transformations to F-distributions exist in a number of cases (under the null hypothesis).

For $h = 1$ and any p,

$$\frac{1-\Lambda}{\Lambda}\frac{(r-p+1)}{p} \sim F(p, r-p+1) \tag{8.19}$$

This is equivalent to the use of an Hotelling T^2-test and it is easy to show that $(1-\Lambda)/\Lambda$ is distributed as $(1/r)T_p^2(r)$ (see Exercise 8.3).

For $h = 2$ and any p,

$$\frac{1-\Lambda^{1/2}}{\Lambda^{1/2}}\frac{(r-p+1)}{p} \sim F(2p, 2(r-p+1)) \tag{8.20}$$

† A better but more complicated approximation due to Rao (1951) is that

$$\frac{1-\Lambda^{1/b}}{\Lambda^{1/b}}\frac{(ab-c)}{ph} \sim F(ph, ab-c)$$

where $a = \left(r - \frac{p-h+1}{2}\right)$, $b = \sqrt{\{(p^2h^2-4)/(p^2+h^2-5)\}}$, $c = \frac{ph-2}{2}$

This reduces to the exact transformations given above for $h = 1$ and 2.

and since $\Lambda_{p,h,r} = \Lambda_{h,p,r+h-p}$ (Anderson, 1958, Theorem 8.4.2), for $p = 2$ and any h,

$$\frac{1-\Lambda^{1/2}}{\Lambda^{1/2}}\frac{(r-1)}{h} \sim F(2h, 2(r-1)) \tag{8.21}$$

For $p = 1$, the statistic reduces to the usual variance ratio statistic.

8.3.5 *Applications of the methodology*

Let us now use the procedure to test the hypothesis of no treatment effects in Example 8.1.

$$\Lambda_{3,2,4} = 0.004\,313.$$

Using Equation (8.18),

$$\left(r - \frac{p-h+1}{2}\right) = 3 \qquad ph = 6$$

$$-3 \ln \Lambda = 16.34 \qquad \chi^2_{.01}(6) = 16.81$$

so that the result is nearly significant at the 1 per cent level ($P = 0.012$).†

Alternatively, using Equation (8.20), since $h = 2$, and an exact transformation is available,

$$\frac{1-\Lambda^{1/2}}{\Lambda^{1/2}}\frac{r-p+1}{p} = 14.2269 \times \frac{2}{3} = 9.48$$

$$F_{.025}(6, 4) = 9.20 \qquad F_{.01}(6, 4) = 15.21$$

giving significance at the 2.5 per cent level ($P = 0.024$). It is apparent that the approximate test, Equation (8.18), is not very good in this application. This is not surprising since it is based on asymptotic theory and here $r = 4$. We shall see later that for larger values of r the approximation is excellent.

A separate univariate analysis on each of the components gives variance ratios 7.30, 7.60 and 3.17 with $F_{.05}(2, 4) = 6.94$. The fact that the multivariate test puts the significance level at 2.5 per cent indicates that, for these data, the evidence provided by the three components is, in some sense, cumulative. This illustrates the purpose of the multivariate test for, without it, we cannot know whether the significance of the second component is merely repeating the significance of the first component (because of the intercorrelation) or whether it is providing additional evidence as shown by a lower significance level in the overall test.

We shall return to the analysis of Example 8.1 in later sections. For the moment let us turn to Example 8.2.

It should be observed that in this example with three-dimensional data, the first component is usually treated as a covariate, the other two components

† The exact significance levels quoted in brackets were obtained using the statistics module programs on a Texas T.I. 58 calculator.

Table 8.7 MANOVA table for Example 8.2

Source	d.f.	SSPM (omitting lower triangle)		
Temperature (T)	2	4.8161	9.6640	.28437
			32.5867	.37693
				.019633
Sex (S)	1	.6422	1.2712	−.19644
			2.5163	−.38884
				.060089
$T \times S$	2	.2755	.8160	.03818
			3.2054	.08818
				.006078
Residual	12	19.3264	7.0094	−.19063
			26.6988	.20847
				.039200
Total	17	25.0602	18.7606	−.06453
			65.0072	.28473
				.125000

forming a bivariate response. We shall adopt this approach in Chapter 9, but in this chapter we shall regard the data as referring to a three-dimensional response. The MANOVA table calculated from the data in Table 8.3 is given to five significant figures in Table 8.7.

To test the hypothesis that there is no interaction, take H to be SSPM $(T \times S)$ on 2 d.f. $|R| = 15.9336$, $|H + R| = 20.6405$, giving $\Lambda_{3,2,12} = 0.7720$. Using Equation (8.20), since $h = 2$,

$$\frac{1 - \Lambda^{1/2}}{\Lambda^{1/2}} \frac{r - p + 1}{p} = 0.1381 \times \frac{10}{3} = 0.46$$

which is obviously not significant.

Alternatively, using Equation (8.18), $r - (p - h + 1)/2 = 11$; $ph = 6$ and $-11 \ln \Lambda = 2.85$, which is also not significant since $\chi^2_{.10}(6) = 10.64$.

Both tests in fact agree on an exact significance level of 83 per cent.

We may now assume that the temperature effects (if any) are the same for both sexes and that the sex effect (if any) is the same at each temperature, and proceed to test the main effects.

To test the hypothesis that there are no temperature effects, take H to be SSPM(T) on 2 d.f. $|R| = 15.9336$ as before, $|H + R| = 60.8862$, giving $\Lambda_{3,2,12} = 0.2617$.

Using Equation (8.20),

$$\frac{1 - \Lambda^{1/2}}{\Lambda^{1/2}} \frac{r - p + 1}{p} = 0.9548 \times \frac{10}{3} = 3.18$$

$$F_{.025}(6, 20) = 3.13 \qquad (P = 0.023)$$

Or, using Equation (8.18),

$$-11\ln\Lambda = 14.75;\ \chi^2_{.025}(6) = 14.45 \qquad (P = 0.022)$$

With a significance level of 2.5 per cent we can reasonably reject the null hypothesis. Examination of the data (Table 8.3 – look at the temperature mean vectors) and of the separate univariate tests seems to indicate that the effect lies in the detrimental effect of low and high temperatures on the weight gain $(X_2 - X_1)$ and in the low tumour weight at low temperature.

To test the hypothesis that there are no sex effects, take H to be SSPM(S) on 1 d.f. $|R| = 15.9336$, as before, $|H + R| = 47.2451$, giving $\Lambda_{3,1,12} = 0.3373$. Using Equation (8.19),

$$\frac{1-\Lambda}{\Lambda}\frac{r-p+1}{p} = 1.9647 \times \frac{10}{3} = 6.55$$

$$F_{.01}(3,10) = 6.55 \qquad (P = 0.01)$$

Or, using Equation (8.18),

$$r - \frac{p-h+1}{2} = 10.5;\ ph = 3$$

$$-10.5\ln\Lambda = 11.41;\ \chi^2_{.01}(3) = 11.34$$
$$(P = 0.0097)$$

We would again reject the null hypothesis. Examination of the data and univariate analyses seems to indicate that the sex difference lies in the tumour weight – males give heavier tumours than females.

Certainly by this stage we would have reached the decision that to obtain reasonably large tumours in a comparatively vigorous control group we should use male mice only and rear them at about 20 °C. The statistical analysis confirms first impressions. However, we will take the analysis further in later sections to illustrate other available techniques.

8.4 Further analysis

Once the null hypothesis is rejected, it is necessary to study in detail the discrepancies between the null hypothesis and the data. It is possible to find confidence regions based on the Wilks' Λ-statistic but, as in the two-sample case (Section 7.6), this approach is of little practical use. Confidence intervals for contrasts between expected values of particular linear compounds could be derived but they would be impractically wide to achieve overall protection against the type 1 error. We might argue, as in Section 7.3, that confidence intervals based on the t-distribution could be used to pick out contrasts of interest. However, the likelihood of spuriously significant results increases with the size of the data set and great caution would have to be observed.

In this chapter we shall introduce a technique known as *canonical variates analysis* (CVA) as an alternative approach. This is essentially a technique for choosing those linear compounds of the multivariate response which demonstrate the greatest inconsistency between null hypothesis and data. It is an extension of the technique introduced in Sections 7.3 and 7.6 which led to discriminant analysis.

8.5 The dimensionality of the alternative hypothesis

Let us introduce this section by referring to Example 8.1, where a three-dimensional response is observed for three treatments. Suppose the expected values for the three treatments are $\boldsymbol{\mu}_i$, $i = 1, 2, 3$, and let them be represented by points in three-dimensional space using orthogonal axes to represent the three components.

Any three points must lie in a plane which defines a two-dimensional subspace. Note that this holds not only for a three-dimensional response but for any p-dimensional response with $p \geq 2$. Now consider the case when the three points are collinear† and so lie in a one-dimensional subspace. We can say that the alternative hypothesis is one-dimensional compared to the general case when it is two-dimensional. Of course, when the three points coincide the null hypothesis is true.

Now consider the equation

$$|H_\mu - \theta \mathrm{r} \Sigma| = 0 \tag{8.22}$$

which is like Equation (8.10) but with H_μ denoting the hypothesis matrix calculated using the $\boldsymbol{\mu}_i$ rather than the treatment mean vectors and with $r\Sigma$ replacing its estimate R on r d.f.

The rank of H_μ and hence the number of non-zero roots of Equation (8.22) will depend on the dimensionality of the $\boldsymbol{\mu}_i$, and conversely the number of non-zero roots defines the dimensionality of the $\boldsymbol{\mu}_i$. Thus two non-zero roots indicate that the three points are neither coincident nor collinear. One non-zero root indicates that the three points are collinear but not coincident. No non-zero roots indicates that the three points are coincident and that the null hypothesis holds.

Let us look briefly at Example 8.3. There are four treatments with a bivariate response. Equation (8.22) can have only two non-zero roots and the dimensionality of the four $\boldsymbol{\mu}_i$ cannot be greater than 2, since points are plotted in two dimensions. If only one root of Equation (8.22) is non-zero, then we would infer that the four $\boldsymbol{\mu}_i$ are collinear.

In general, the maximum dimensionality of the alternative hypothesis is $s = \min(h, p)$ and the actual dimensionality is the number of non-zero roots of Equation (8.22). We need not restrict application to treatment means. For instance, if interaction had been found significant in Example 8.2 it would be natural to compare 'cell' means as opposed to 'row' and 'column' means, but it

† This includes, for example, the case when $\boldsymbol{\mu}_1 = \boldsymbol{\mu}_2 \neq \boldsymbol{\mu}_3$.

would also be possible to examine the dimensionality of the interaction parameters by taking H to be the interaction matrix $T \times S$.

In practice the roots are estimated from Equation (8.10) with $\theta_1 \geq \theta_2 \geq \ldots \geq \theta_p$, and a test of hypothesis is required to determine the number of roots to be assumed non-zero. To test the hypothesis that the dimensionality is $t < s$, it can be shown that the likelihood-ratio approach leads to the intuitively satisfactory test statistic $r(\theta_{t+1} + \theta_{t+2} + \ldots + \theta_p)$. A derivation of the result may be found in Rao (1973, Section 8c.6), who also shows that the statistic is asymptotically distributed as $\chi^2[(p-t)(h-t)]$ under the null hypothesis.

However, we shall follow Bartlett (1947), who formulates the statistic in terms of λ_i, the roots of Equation (8.14), where $\lambda_i = (1+\theta_i)^{-1}$. Bartlett assumes that, under the null hypothesis,

$$-\left(r - \frac{p-h+1}{2}\right) \sum_{i=t+1}^{p} \ln \lambda_i \sim \chi^2[(p-t)(h-t)] \tag{8.23}$$

Note that this is essentially the same statistic, since for small θ_i, $-\ln \lambda_i = \ln(1+\theta_i) \simeq \theta_i$.

When $t = 0$, Equation (8.23) reduces to Equation (8.18), the overall Wilks' Λ-test for the hypothesis matrix H. We shall find it convenient to calculate the statistic as

$$\left(r - \frac{p-h+1}{2}\right) \ln\left\{\prod_{i=t+1}^{p} (1+\theta_i)\right\}$$

since, using a calculator, it is easier to solve Equation (8.10) for the θ_i than to solve Equation (8.14) for the λ_i.

In Example 8.1, let us test the hypothesis that the means are collinear. We found previously that $\theta_1 = 145.613$, $\theta_2 = 0.581535$, and we note that θ_2 is very small compared with θ_1. The test statistic is

$$3 \ln(1.581535) = 1.38; \ (p-t)(h-t) = 2$$
$$\chi^2_{.05}(2) = 5.99$$

θ_2 is not significantly different from zero and we may therefore infer that the means are collinear.

Another example is given in Section 8.6.

8.6 Canonical variates analysis

Having decided that the expected values with which we are concerned lie in a subspace of dimension t, we proceed to compare the sample means in the estimated subspace. To do this we require to find the co-ordinates of the sample means in this subspace relative to axes which we have to choose. These axes will represent linear compounds of the response vector.

Consider $Z = \mathbf{a}^T\mathbf{X}$, a linear compound of the response vector $\mathbf{X}$. By an application of Equation (6.21), the univariate ANOVA table for Z may be obtained from the MANOVA table by calculating the sum of squares for each source as $\mathbf{a}^T \text{SSPM}\,\mathbf{a}$. For a hypothesis matrix H on h d.f., the variance ratio statistic for testing the corresponding univariate hypothesis will be:

$$\text{V.R.} = \frac{(\mathbf{a}^T H\mathbf{a})/h}{(\mathbf{a}^T R\mathbf{a})/r} \tag{8.24}$$

We shall choose Z_1, the first *canonical variate*, as that linear compound for which the variance ratio is maximum. Z_2, the second canonical variate, will be chosen as that linear compound which maximizes the variance ratio subject to the constraint $\hat{\text{Cov}}(Z_1, Z_2) = 0$. This process will be continued until all t canonical variates are defined. When the degrees of freedom for H is one, this is essentially the derivation of the Hotelling T^2-statistic for two samples and, as in that case, $\mathbf{a}$ is not uniquely determined since the variance ratio is unaltered if $\mathbf{a}$ is replaced by $k\mathbf{a}$, where k is a scalar. To obtain a unique solution we add the constraint

$$\mathbf{a}^T R\mathbf{a} = r \tag{8.25}$$

chosen so that

$$\hat{\text{Var}}(\mathbf{a}^T\mathbf{X}) = \mathbf{a}^T\hat{\Sigma}\,\mathbf{a} = \mathbf{a}^T(R/r)\mathbf{a} = 1 \tag{8.26}$$

The problem may be written as:
Find $\mathbf{a}$ so as to maximize $\mathbf{a}^T H\mathbf{a}$ subject to $\mathbf{a}^T R\mathbf{a} = r$.

The Lagrangian is

$$L_1(\mathbf{a}) = \mathbf{a}^T H\mathbf{a} + \theta(r - \mathbf{a}^T R\mathbf{a}) \tag{8.27}$$

Then $\mathrm{d}L_1/\mathrm{d}\mathbf{a} = 2H\mathbf{a} - 2\theta R\mathbf{a}$ (see Equation (1.18))

For stationarity,

$$(H - \theta R)\mathbf{a} = \mathbf{0} \tag{8.28}$$

A solution exists if $|H - \theta R| = 0$. Thus we meet Equation (8.10) again.

If θ is a root of Equation (8.10), we find the corresponding vector $\mathbf{a}$ by solving Equation (8.28) subject to the constraint (8.25).
Premultiplying Equation (8.28) by $\mathbf{a}^T$, we obtain

$$\mathbf{a}^T H\mathbf{a} = \theta\mathbf{a}^T R\mathbf{a} = \theta r$$

so that the maximum value is obtained by choosing the largest root θ_1 and its corresponding vector $\mathbf{a}_1$.

We can find the second canonical variate by using the Lagrangian

$$L_2(\mathbf{a}) = \mathbf{a}^T H\mathbf{a} + \theta(r - \mathbf{a}^T R\mathbf{a}) - \phi\mathbf{a}_1^T R\mathbf{a} \tag{8.29}$$

which incorporates the constraint $\hat{\text{Cov}}(Z_1, Z) = \mathbf{a}_1^T R\mathbf{a}/r = 0$.

This is left as an exercise to the reader, who can also extend the procedure to find the other canonical variates.

We shall instead turn to an alternative derivation which depends on a result concerning the roots of

$$(H - \theta R)\mathbf{v} = \mathbf{0} \tag{8.30}$$

Since R is a real symmetric matrix of full rank, $R^{1/2}$ exists and Equation (8.30) may be rewritten as

$$R^{1/2}(R^{-1/2}HR^{-1/2} - \theta I)R^{1/2}\mathbf{v} = \mathbf{0} \tag{8.31}$$

where $R^{-1/2}HR^{-1/2}$ is also symmetric (but not necessarily of full rank). Pre-multiplying by $R^{-1/2}$, we see that Equation (8.31) is of the form $(A - \lambda I)\mathbf{c} = \mathbf{0}$ of Section 1.6 and, if $R^{1/2}(\mathbf{v}_1, \mathbf{v}_2, \ldots, \mathbf{v}_p) = R^{1/2}V$ is the matrix of eigenvectors† and Θ the diagonal matrix of eigenvalues, we have by Equation (1.15),

$$(R^{1/2}V)^{\mathrm{T}}(R^{-1/2}HR^{-1/2})(R^{1/2}V) = V^{\mathrm{T}}HV = \Theta \tag{8.32}$$

and

$$(R^{1/2}V)^{\mathrm{T}}(R^{1/2}V) = V^{\mathrm{T}}RV = I \tag{8.33}$$

Now V is of full rank so that, for any $(p \times 1)$ vector $\mathbf{a}_i$, there exists a unique $\mathbf{k}_i$ such that

$$\mathbf{a}_i = V\mathbf{k}_i = \sum_{r=1}^{p} k_{ir}\mathbf{v}_r \tag{8.34}$$

Note especially from this and Equation (8.33) that

$$\mathbf{v}_j^{\mathrm{T}}R\mathbf{a}_i = 0 \text{ implies } k_{ij} = 0 \tag{8.35}$$

Hence for any $\mathbf{a}_i$,

$$\frac{\mathbf{a}_i^{\mathrm{T}}H\mathbf{a}_i}{\mathbf{a}_i^{\mathrm{T}}R\mathbf{a}_i} = \frac{\mathbf{k}_i^{\mathrm{T}}V^{\mathrm{T}}HV\mathbf{k}_i}{\mathbf{k}_i^{\mathrm{T}}V^{\mathrm{T}}RV\mathbf{k}_i} = \frac{\mathbf{k}_i^{\mathrm{T}}\Theta\mathbf{k}_i}{\mathbf{k}_i^{\mathrm{T}}\mathbf{k}_i} = \frac{k_{i1}^2\theta_1 + k_{i2}^2\theta_2 + \ldots + k_{ip}^2\theta_p}{k_{i1}^2 + k_{i2}^2 + \ldots + k_{ip}^2} \tag{8.36}$$

We may now return to the problem of finding the maximum of the variance ratio in Equation (8.24). From Equation (8.36) it is clear that the variance ratio achieves a maximum value $r\theta_1/h$ when $k_{12} = k_{13} = \ldots = k_{1p} = 0$ so that $\mathbf{a} = \mathbf{a}_1 = k_{11}\mathbf{v}_1$. k_{11} is chosen so that $\mathbf{a}_1^{\mathrm{T}}R\mathbf{a}_1 = r$ to satisfy Equation (8.25).

To obtain the second canonical variate, note that the requirement that $\hat{\mathrm{Cov}}(Z_1, Z_2) = 0 \Rightarrow k_{21} = 0$, by Equation (8.35). The maximum of $(\mathbf{a}_2^{\mathrm{T}}H\mathbf{a}_2/\mathbf{a}_2^{\mathrm{T}}R\mathbf{a}_2)$ subject to $k_{21} = 0$ is clearly θ_2 (corresponding to a variance ratio of $r\theta_2/h$ achieved when $\mathbf{a}_2$ is proportional to $\mathbf{v}_2$).

Hence Equation (8.36) establishes that the canonical variates are proportional to $\mathbf{v}_i^{\mathrm{T}}\mathbf{X}$, $i = 1, 2, \ldots, t$, with corresponding variance ratios $r\theta_1/h$ and we see that the canonical variates are ordered in terms of the variance ratio.

When H is of full rank, so that $\theta_p > 0$, the minimum of Equation (8.24) is $r\theta_p/h$, achieved when $\mathbf{a} \propto \mathbf{v}_p$. The result will be used in Section 8.8.

Equation (8.31) points to a similarity in the derivation of canonical variates

† Since there may be a multiplicity of (at least) the zero root, $R^{1/2}V$ is not necessarily unique. The result is true for any suitable choice of eigenvectors.

and principal components. Seal (1964) uses this similarity to derive the canonical variates and gives some useful examples. However, unlike a principal components analysis, canonical variates analysis (CVA) is *invariant under change of origin and scale of measurement.*

In Example 8.1 we found that we could assume that only one dimension was required for the further examination of the means.

Suppose $\mathbf{a}_1^{\mathrm{T}}$ is proportional to $[a, b, 1]$, since we can solve Equation (8.28) only for the ratios of the elements of $\mathbf{a}_1$. Then Equation (8.28) becomes

$$\begin{bmatrix} (1.68 - 0.46\theta_1) & (1.38 - 0.03\theta_1) & (-1.26 + 0.40\theta_1) \\ (1.38 - 0.03\theta_1) & (1.14 - 0.30\theta_1) & (-1.08 + 0.48\theta_1) \\ (-1.26 + 0.40\theta_1) & (-1.08 + 0.48\theta_1) & (1.26 - 1.06\theta_1) \end{bmatrix} \begin{bmatrix} a \\ b \\ 1 \end{bmatrix} = \mathbf{0}$$

giving, with $\theta_1 = 145.613$,

$$\begin{aligned} 65.302a + 2.988b &= 56.985 \\ 2.998a + 42.544b &= 68.814 \\ 56.985a + 68.814b &= 153.090 \end{aligned}$$

The third equation is redundant but may be used for checking. Solving the first two equations, we obtain

$$\mathbf{a}_1^{\mathrm{T}} = c[0.801, 1.561, 1]\dagger$$

The scaling constant c is obtained from Equation (8.25):

$$\mathbf{a}_1^{\mathrm{T}} R \mathbf{a}_1 = 0.0218c^2 = 4$$

Hence $c = 13.54$ and $\mathbf{a}_1^{\mathrm{T}} = [10.85, 21.14, 13.54]$

Check:
$$\hat{\mathrm{Var}}\, Z_1 = \mathbf{a}_1^{\mathrm{T}} R \mathbf{a}_1/4 = 1.001$$
$$\mathbf{a}_1^{\mathrm{T}} H \mathbf{a}_1 = 582.816$$

so that the variance ratio for Z_1 is $291.408/1.001 = 291.12$, which is in reasonable agreement with $2\theta_1 = 291.23$.

We now calculate $\bar{z}_{1i}$, $i = 1, 2, 3$, for the treatment means using $\bar{z}_{1i} = \mathbf{a}_1^{\mathrm{T}} \bar{\mathbf{x}}_i$, and obtain 637.61, 629.26, 648.90. Since the standard error for each mean is $\sqrt{(1/3)} = 0.58$ (why?), we are in no doubt that the treatment means are different. They are approximately equally spaced and ordered as $\bar{z}_{12} < \bar{z}_{11} < \bar{z}_{13}$.

Another check is available by calculating the sum of squares for treatments directly from the above means, obtaining 582.92, in good agreement with the previous calculation.

We stated earlier that CVA was invariant under change of scale. Let us see how this statement is reflected in the equation for the first component we have obtained. Suppose the values of X_3 in Table 8.1 had been scaled by

† Substituting in the check equation we find LHS = 153.064. We can achieve a better agreement by keeping more decimal places in the calculations but the apparent accuracy is spurious. The third (and even the second) decimal place given above is suspect and is only retained for intermediate calculations.

multiplying by a constant k, say. Then by following through the calculations above, it is easy to see that the equation for the first canonical variate would be

$$Z_1 = 10.85X_1 + 21.14X_2 + \left(\frac{13.54}{k}\right)(kX_3)$$

giving exactly the same values to Z_1 as before. Note also that the contribution to the values of Z_1 from the third component of the response is exactly the same no matter the scaling. However, the value of the coefficient a_3 does depend on the scaling. If we wish to attempt to describe what the canonical variate is measuring or to assess the relative importance of the components of the response in the canonical variate, we have to take account of the scales of measurement.One way to do this might be to standardize the components of the response by dividing by the respective standard deviations (estimated from the residual matrix R). Here we would obtain

$$\begin{aligned} Z_1 &= 10.85X_1 + 21.14X_2 + 13.54X_3 \\ &= 3.68\left(\frac{X_1}{s_1}\right) + 5.79\left(\frac{X_2}{s_2}\right) + 6.97\left(\frac{X_3}{s_3}\right) \end{aligned}$$

since s_1, for example, is given by $\sqrt{(0.46/4)} = 0.339$. This equation indicates here that, when the components are measured with the same precision, X_3 seems to play a slightly more important part than X_2 and both of these are more important than X_1 in discriminating among the treatments.

The second canonical variate, corresponding to $\theta_2 = 0.5815$, is

$$Z_2 = 0.57X_1 + 1.56X_2 + 2.86X_3$$

giving treatment means 82.90, 81.77 and 81.87 with a variance ratio of 1.17 ($2\theta_2 = 1.16$). In the previous section we found that θ_2 was not significantly different from zero. However, we might still wish to find the corresponding canonical variate, since when $\theta = 0$ the corresponding canonical variate is a linear compound which is constant for the different treatments (the variance ratio is zero).

As an exercise, estimate the block effect for Z_2 and remove it from the data. Note how similar the resultant values are, remembering that $\hat{\text{Var}}\, Z_2 = 1$. The treatment means not being significantly different, the best estimate of the common mean is 82.18, the grand mean of the Z_2 values. We may think of Z_2 as giving a linear relationship applicable to all three treatments among the components of the mean vector estimated by

$$0.57\mu_1 + 1.56\mu_2 + 2.86\mu_3 = 82.18$$

We shall return briefly to this topic in Section 8.7.

In Example 8.2 the temperature effect can be similarly treated and will be left as an exercise. If interaction had been found significant, the obvious approach would be to carry out a canonical variates analysis on the cell means. H is taken to be the SSPM for treatments calculated as $H = T + S + T \times S$ (see Table 8.7) on 5 d.f.; $s = \min(p, h) = 3$

The three non-zero roots are 2.6800, 1.4080 and 0.1095 and $\Lambda_{3,5,12} = 0.1017$. Note that here, in a fully randomized experiment, $H + R =$ SSPM(Total).

The application of the testing procedure in Section 8.5 gives

H_0:Dimensionality = 2. $12.5 \ln(1.1095) = 1.30$; $\chi^2_{.05}(3) = 7.82$

H_0:Dimensionality = 1. $12.5 \ln(2.4080 \times 1.1095) = 12.28$; $\chi^2_{.05}(8) = 15.51$

H_0:Dimensionality = 0. $12.5 \ln(3.6800 \times 2.4080 \times 1.095) = 28.57$; $\chi^2_{.05}(15) = 25.00$

The last of these tests is the overall Wilks Λ-test.

The non-significance of the test that dimensionality = 1 could be due to the fact that we have included the interaction term in the treatments. This has already been shown to be non-significant and its inclusion has diluted the main effects. We shall, however, proceed as though we had accepted that a dimensionality of 2 was appropriate. (Repeat the analysis with $H = T + S$ on 3 d.f.)

Assuming that θ_1 and θ_2 are non-zero, the canonical variates are

$$Z_1 = 0.306X_1 - 0.333X_2 + 18.474X_3$$

and

$$Z_2 = 0.107X_1 + 0.609X_2 + 2.938X_3$$

giving the co-ordinates of the treatment means shown in Table 8.8.

The means may be plotted as in Fig. 8.1, where 1, 2 and 3 denote the temperature levels 4 °C, 20 °C and 34 °C respectively.

Since the canonical variates are normalized so that $\hat{\text{Var}}(Z) = 1$, the standard errors of the means are simply $1/\sqrt{n}$. Hence, since the Z's are uncorrelated, approximate confidence regions for the (true) means may be indicated by circles centred on the sample means with a radius of $k/\sqrt{n}$, where in this case $n = 3$, $k = 1$ would give approximate 68% confidence regions, $k = 2$ approximate 95% confidence regions and so on. The plot, with $k = 1$, shows clearly that there are sex and temperature effects. There is, however, a suggestion of the presence of interaction, in contradiction to the result of the hypothesis testing in Section 8.3, since the plot seems to indicate that temperature changes affect males more strongly than females.

Table 8.8 Two-dimensional co-ordinates of treatment means

Sex	Temperature (°C)		
	4	20	34
M	4.24, 14.10	6.02, 16.39	6.32, 13.58
F	2.81, 14.54	3.20, 15.69	3.77, 14.30

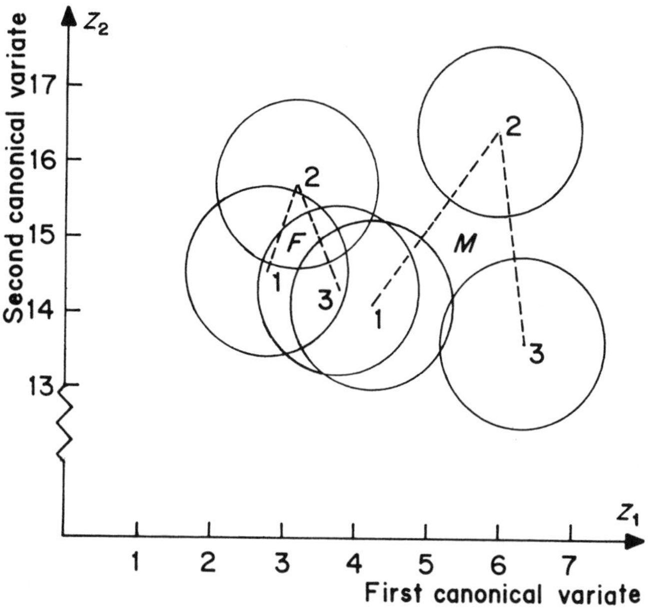

Figure 8.1. Canonical variates plot of the six treatment means using the first two canonical variates.

In cases where p is large, CVA may be used to provide a low-dimensional approximation to the data. We might know that the true dimensionality is large but choose to examine a two-dimensional approximation. CVA produces a best two-dimensional approximation in the least-squares sense.

8.7 Linear functional relationships

Suppose k populations have means $\boldsymbol{\mu}_i$, $i = 1, 2, \ldots, k$, and we are interested in determining whether or not there exists a linear relationship of the form.

$$a_1\mu_{r1} + a_2\mu_{r2} + \ldots + a_p\mu_{rp} = \text{constant} \tag{8.37}$$

which is satisfied by the $\boldsymbol{\mu}_i$.

First of all, consider the case where $k - 1 > p$. Then Equation (8.37) implies that the $\boldsymbol{\mu}_i$ lie in a subspace of dimension $(p - 1)$ and, as discussed in Section 8.5, the equation $|H_\mu - \theta r\Sigma| = 0$ has a zero root, with corresponding vector $\mathbf{a}^T = [a_1, a_2, \ldots, a_p]$. The existence of more than one linear relationship corresponds to the existence of more than one zero root.

The estimation procedure is exactly that of CVA but with the focus of attention on the canonical variates corresponding to those roots θ_i which are not significantly different from zero. This application is discussed in detail in Sprent (1969) and the derivation of the second canonical variate for Example 8.1 in the previous section provides an illustration.

On the other hand, if $k - 1 < p$, so that $\text{rank}(H) = k - 1$, the k means necessarily lie in a space of dimension $k - 1 < p$. There is no suggestion that, if a further population were to be brought into the analysis, its mean would lie in this $(k - 1)$-dimensional space. However, if the k means had been found to lie in a subspace of dimensionality $t < k - 1$, we could infer the existence of a linear functional relationship.

8.8 Discriminant analysis

In Section 8.6, CVA was introduced as a technique for examining the differences among the means of a number of populations. When only two populations are involved, the first (and only) canonical variate is just the linear discriminant function discussed in Section 7.8. In that section it was suggested that when there are more than two populations they could be compared two at a time, and the discrimination procedure developed was based on equivalent discriminant scores. Alternatively, suppose a CVA is performed on data known to arise from $k > 2$ populations to produce a set of co-ordinates for the means in a subspace of minimum dimension. The values of the co-ordinates for any unassigned individual may then be similarly calculated and the individual allocated to that population with the nearest mean. Hence we see that CVA provides an overall allocation procedure in a dimensionality determined by the dimensionality of the population means. The method based on equivalent discriminant scores is essentially a pairwise procedure, although each pair of populations would be compared in a different single dimension. The two techniques would coincide if the population means were collinear.

Exercises

8.1 (i) Calculate the MANOVA table for Example 8.3 (Table 8.4).
(ii) Partition SSPM (Nitrogen) into linear and non-linear components.
(iii) Calculate the Λ-statistic for testing: (*a*) the overall treatment effect; (*b*) the non-linear nitrogen effect; (*c*) the linear nitrogen effect. Test the significance of the effects using the χ^2- and F-approximations given by Equations (8.18)–(8.21).

8.2 Using the χ^2-approximation in Exercise 8.1(iii), we observe that the partitioning is not additive in terms of the χ^2 values. Bartlett (1947) suggests the following:

$$\text{Write} \quad \frac{|R|}{|N+R|} = \frac{|R|}{|\text{non-lin. } N+R|} \cdot \frac{|\text{non-lin. } N+R|}{|N+R|}$$

Take natural logs and multiply throughout by $-[r-(p-h+1)/2]$, with $h = 3$ the d.f. for $N = \text{SSPM (Nitrogen)}$. The partition is now additive. Comment on this approach.

8.3 (i) Two treatments are compared in terms of a p-variate response in a fully randomized experiment. If $\mathbf{D}$ is the $(p \times 1)$ vector difference of the two treatment mean vectors, show that the SSPM(Treatments) on 1 d.f. in a MANOVA may be written as $[n_1n_2/(n_1+n_2)]\mathbf{DD}^{\mathrm{T}}$, where n_i, $i = 1, 2$, is the number of observations on the ith treatment.

(ii) Using the result for partitioned matrices that

$$\begin{vmatrix} K & L \\ M & N \end{vmatrix} = |K||N - MK^{-1}L| = |N||K - LN^{-1}M|$$

with $N = -1$, $L = M^{\mathrm{T}} = [n_1n_2/(n_1+n_2)]^{1/2}\mathbf{D}$, $K = R$ the residual matrix on $r = n_1 + n_2 - 2$ d.f., show that the Λ statistic and the Hotelling T^2-statistic are related by

$$\frac{1}{\Lambda} = 1 + \frac{\mathcal{T}^2}{r}$$

8.4 Two treatments are compared in terms of a p-variate response in a randomized block experiment. Establish the connection between the Λ-statistic in a MANOVA with the Hotelling T^2-statistic for a paired comparison test.

8.5 (i) With $H =$ SSPM (Nitrogen) in Example 8.3, carry out a CVA.

(ii) Estimate a linear relationship applying to the components of the treatment means.

(iii) Use CVA to find the linear compound of the response which has the best linear relationship with the nitrogen levels and estimate the relationship.

(iv) The second root of $|H - \theta R| = 0$ in part (iii) above is zero. Show that the corresponding canonical variate expresses an exact linear relationship between the estimates of the treatments obtained using the fitted values of the linear relationship with nitrogen levels derived in part (iii).

8.6 Example 8.2 may be tackled by analysing the bivariate response $\mathbf{U}$, where $U_1 = Y_2 - Y_1$, $U_2 = Y_3$. Carry out this analysis.
[*Hint*: $\mathbf{U} = C^{\mathrm{T}}\mathbf{Y}$, where

$$C^{\mathrm{T}} = \begin{bmatrix} -1 & 1 & 0 \\ 0 & 0 & 1 \end{bmatrix}$$

and the U-MANOVA may be obtained from the Y-MANOVA by $\mathrm{SSPM}_U = C^{\mathrm{T}}\mathrm{SSPM}_Y C$.]

CHAPTER NINE

The multivariate analysis of covariance and related topics

9.1 Introduction

In Chapter 8 we discussed the methodology for analysing experiments with a p-variate response, the methodology being based on multivariate normal-distribution theory. In this chapter we will be concerned with further methodology also based on normal theory but with the emphasis on the conditional distribution of a subset of the response variates given the values of the others. The techniques to be discussed are multivariate regression, multivariate analysis of covariance (MANOCOVA) and the test for additional information. The approach will again be heuristic with the procedures being presented as generalizations of univariate procedures.

9.2 Multivariate regression

Suppose the p-variate vector $\mathbf{X}$ is partitioned into a $(q \times 1)$ vector $\mathbf{X}_1$ and a $((p-q) \times 1)$ vector $\mathbf{X}_2$. Then, if $\mathbf{X} \sim N_p(\boldsymbol{\mu}, \Sigma)$, the conditional distribution of $\mathbf{X}_1$ given $\mathbf{X}_2 = \mathbf{x}_2$ is normal with

$$E(\mathbf{X}_1 | \mathbf{X}_2 = \mathbf{x}_2) = \boldsymbol{\mu}_1 + \Sigma_{12}\Sigma_{22}^{-1}(\mathbf{x}_2 - \boldsymbol{\mu}_2) \tag{9.1}$$

and

$$\mathrm{Var}(\mathbf{X}_1 | \mathbf{X}_2 = \mathbf{x}_2) = \Sigma_{11} - \Sigma_{12}\Sigma_{22}^{-1}\Sigma_{21} = \Sigma_{q.(p-q)} \tag{9.2}$$

using the results and notation of Section 6.3 (iv) and assuming the existence of Σ_{22}^{-1}.

The abbreviated form of the conditional covariance matrix $\Sigma_{q.(p-q)}$ introduces a notation which will be used throughout the chapter. In this notation, the first figure in the subscript refers to the number of components in the dependent variate, $\mathbf{X}_1$, and the second to the number in the independent variate or regressor, $\mathbf{X}_2$. The elements of $\Sigma_{q.(p-q)}$ will be denoted by $\sigma_{ij.(p-q)}$.

By an extension of the definition of a regression function of a univariate dependent variable, Equation (9.1) may be called the regression function of $\mathbf{X}_1$ on $\mathbf{X}_2$. Then $\Sigma_{12}\Sigma_{22}^{-1}$ is the matrix of regression coefficients and Equation (9.2) gives the variance about the regression.

Multivariate regression is the technique used to estimate the regression coefficients and the variance about the regression from a given $(n \times p)$ data matrix X containing n independent observations on $\mathbf{X}$.

Suppose X is written as

$$X = [X_1, X_2] = [[\mathbf{y}_1, \mathbf{y}_2, \ldots \mathbf{y}_q][\mathbf{z}_{(q+1)}, \mathbf{z}_{(q+2)}, \ldots \mathbf{z}_p]] \tag{9.3}$$

where the $(n \times 1)$ column vector $\mathbf{y}_j$ denotes the n independent observations on the jth component of $\mathbf{X}_1$ (and of $\mathbf{X}$) and the $(n \times 1)$ column vector $\mathbf{z}_{(q+k)}$ denotes the values taken by the kth component of $\mathbf{X}_2$ (and $(q+k)$th component of $\mathbf{X}$) on which the distribution of $\mathbf{X}_1$ is to be conditioned. This implies that X_2 is to be regarded as a matrix of given constants.

Let the estimates $\hat{\boldsymbol{\mu}} = \bar{\mathbf{x}}$ and $\hat{\Sigma} = S$ be correspondingly partitioned as

$$\bar{\mathbf{x}} = \begin{bmatrix} \bar{\mathbf{x}}_1 \\ \bar{\mathbf{x}}_2 \end{bmatrix}; \quad S = \begin{bmatrix} S_{11} & S_{12} \\ S_{21} & S_{22} \end{bmatrix} \text{ on } f = (n-1) \text{ d.f.} \tag{9.4}$$

and note that $\bar{\mathbf{x}}_1^{\mathrm{T}} = [\bar{y}_{.1}, \bar{y}_{.2}, \ldots, \bar{y}_{.q}]$, $\bar{\mathbf{x}}_2^{\mathrm{T}} = [\bar{z}_{.(q+1)}, \bar{z}_{.(q+2)}, \ldots, \bar{z}_{.p}]$ and that the matrix of sums of squares and product deviations is simply $(n-1)S$ (see Equation (3.4)).

9.2.1 *The special case: Univariate multiple regression*

Let us begin by considering the special case of a single dependent variable (i.e., $q = 1$). The problem then reduces to the usual multiple-regression model (see Equations (6.6)–(6.8)). This should already be familiar to the reader, but by setting up a suitable notation we can more easily extend the ideas to the case $q > 1$.

We will write the multiple regression in terms of the general linear model of Section 1.7 with $\mathbf{Y}_1$ as the vector random variable giving rise to the observation $\mathbf{y}_1$. Note that $X_1 = \mathbf{y}_1$ in this case and $X_2 = [\mathbf{z}_2, \mathbf{z}_3, \ldots, \mathbf{z}_p]$. Thus we have

$$\Omega : E(\mathbf{Y}_1 | X_2) = \mathbf{1}\beta_0 + G\boldsymbol{\beta}_1 \tag{9.5}$$

where $G = (X_2 - \mathbf{1}\bar{\mathbf{x}}_2^{\mathrm{T}})$, and $\boldsymbol{\beta}_1^{\mathrm{T}} = [\beta_{12}, \beta_{13}, \ldots, \beta_{1p}]$ contains the $(p-1)$ regression coefficients. The observations on the dependent variable are assumed uncorrelated and each has (conditional) variance $\sigma_{11.(p-1)}$ in our notation, so that

$$\operatorname{Var}(\mathbf{Y}_1 | X_2) = \sigma_{11.(p-1)} I_n$$

The parameters are estimated by least squares, thus:

$$\hat{\beta}_0 = \bar{y}_{.1} \tag{9.6}$$

$$\begin{aligned}(X_2 - \mathbf{1}\bar{\mathbf{x}}_2^{\mathrm{T}})^{\mathrm{T}}(X_2 - \mathbf{1}\bar{\mathbf{x}}_2^{\mathrm{T}})\hat{\boldsymbol{\beta}}_1 &= (X_2 - \mathbf{1}\bar{\mathbf{x}}_2^{\mathrm{T}})^{\mathrm{T}}(\mathbf{y}_1 - \mathbf{1}\bar{y}_{.1}) \\ &= (X_2 - \mathbf{1}\bar{\mathbf{x}}_2^{\mathrm{T}})^{\mathrm{T}}(X_1 - \mathbf{1}\bar{\mathbf{x}}_1^{\mathrm{T}})\end{aligned} \tag{9.7}$$

where X_1 is $(n \times 1)$ and X_2 is $(n \times (p-1))$.

Equation (9.7), giving the normal equations for the regression coefficients, involves the partitions of the matrix of corrected sums of square and products. In the notation of Equation (9.4) it may be written as

$$(n-1)S_{22}\hat{\boldsymbol{\beta}}_1 = (n-1)S_{21}$$

On dividing by $(n-1)$, Equation (9.7) may be written in terms of the partitions of S as

$$S_{22}\hat{\boldsymbol{\beta}}_1 = S_{21} \tag{9.8}$$

where, in this univariate case, S_{21} is of order $((p-1)\times 1)$.

The vector of fitted values is given by

$$\hat{E}(\mathbf{Y}_1|X_2) = \mathbf{1}\bar{y}_{.1} + (X_2 - \mathbf{1}\bar{\mathbf{x}}_2^{\mathrm{T}})\hat{\boldsymbol{\beta}}_1 \tag{9.9}$$

and the (uncorrected) sum of squares of the fitted values is

$$\mathrm{SS}(\Omega) = n\bar{y}_{.1}^2 + (n-1)\hat{\boldsymbol{\beta}}_1^{\mathrm{T}}S_{22}\hat{\boldsymbol{\beta}}_1$$

so that, using Equation (9.8),

$$\mathrm{SS}(\Omega) = n\bar{y}_{.1}^2 + (n-1)S_{12}S_{22}^{-1}S_{21} \tag{9.10}$$

where S_{22}^{-1} is assumed to exist.

The residual sum of squares is

$$\begin{aligned} R(\Omega) &= \mathbf{y}_1^{\mathrm{T}}\mathbf{y}_1 - \mathrm{SS}(\Omega) \\ &= \sum_r (y_{r1} - \bar{y}_{.1})^2 - (n-1)S_{12}S_{22}^{-1}S_{21} \\ &= (n-1)(S_{11} - S_{12}S_{22}^{-1}S_{21}) \end{aligned} \tag{9.11}$$

since for $q=1$, $S_{11} = \left[\sum_r (y_{r1} - \bar{y}_{.1})^2\right] \Big/ (n-1)$.

In the abbreviated notation we write $R(\Omega) = (n-1)S_{1.(p-1)}$. The d.f. for $R(\Omega) = n-1-(p-1) = n-p$, since 1 d.f. has been used for the estimation of β_0 and $(p-1)$ d.f. used for the estimation of the β_{1j}, $j = 2, 3, \ldots, p$.

An unbiased estimate for the variance about the regression is given by

$$\hat{\sigma}_{11.(p-1)} = R(\Omega)/(n-p) \tag{9.12}$$

Under the null hypothesis $H_0\colon \boldsymbol{\beta}_1 = \mathbf{0}$, the model Ω reduces to

$$\Omega_0 : E(\mathbf{Y}_1|X_2) = \mathbf{1}\beta_0$$

and the residual sum of squares is simply $(n-1)S_{11}$ on $(n-1)$ d.f.

The sum of squares attributable to the hypothesis, called the sum of squares for regression, is then

$$\mathrm{SS(Regression)} = R(\Omega_0) - R(\Omega) = (n-1)S_{12}S_{22}^{-1}S_{21} \tag{9.13}$$

on $(p-1)$ d.f.

This last expression is a generalization of the well-known result in simple linear regression, for if we regress a single dependent variable Y on a single regressor Z, the sum of squares for regression is

$$\left[\sum_r (z_r - \bar{z}.)(y_r - \bar{y}.)\right]^2 \Big/ \left[\sum_r (z_r - \bar{z}.)^2\right]$$

and this may be written as

$$(n-1)(\hat{\sigma}_{zy})^2/\hat{\sigma}_{zz} = (n-1)\hat{\sigma}_{yz}\hat{\sigma}_{zz}^{-1}\hat{\sigma}_{zy}$$

Finally, let us recall the result from univariate regression theory, Equation (1.24), that

$$\begin{aligned}\operatorname{Var}(\hat{\boldsymbol{\beta}}_1) &= \sigma_{11.(p-1)}[(X_2 - \mathbf{1}\bar{\mathbf{x}}_2^{\mathrm{T}})^{\mathrm{T}}(X_2 - \mathbf{1}\bar{\mathbf{x}}_2^{\mathrm{T}})]^{-1} \\ &= \sigma_{11.(p-1)}S_{22}^{-1}/(n-1)\end{aligned} \tag{9.14}$$

9.2.2 *The general case: Multivariate regression*

Returning now to the general case with $1 \leq q < p$, we may write the model as:

$$\Omega : E(\mathbf{Y}_1, \mathbf{Y}_2, \ldots, \mathbf{Y}_q) = E(X_1) = \mathbf{1}\boldsymbol{\mu}_1^{\mathrm{T}} + GB \tag{9.15}$$

where $G = [X_2 - \mathbf{1}\bar{\mathbf{x}}_2^{\mathrm{T}}]$ and $B = [\boldsymbol{\beta}_1, \boldsymbol{\beta}_2, \ldots, \boldsymbol{\beta}_q]$, with $\boldsymbol{\beta}_j^{\mathrm{T}} = [\beta_{j.(q+1)}, \beta_{j.(q+2)}, \ldots, \beta_{j.p}]$, $j = 1, 2, \ldots, q$

The variance structure is specified by

$$\operatorname{Cov}(\mathbf{Y}_j, \mathbf{Y}_k) = \sigma_{jk.(p-q)}I_n; \quad j, k = 1, 2, \ldots, q \tag{9.16}$$

since it must incorporate the covariance between the jth and kth variates in the same vector observation. $\sigma_{jk.(p-q)}$ is the (j, k)th element of $\Sigma_{q.(p-q)}$.

The least squares estimates of the parameters are given by

$$\hat{\boldsymbol{\mu}}_1 = \bar{\mathbf{x}}_1 \tag{9.17}$$

$$(X_2 - \mathbf{1}\bar{\mathbf{x}}_2^{\mathrm{T}})^{\mathrm{T}}(X_2 - \mathbf{1}\bar{\mathbf{x}}_2^{\mathrm{T}})\hat{B} = (\mathrm{X}_2 - \mathbf{1}\bar{\mathbf{x}}_2^{\mathrm{T}})^{\mathrm{T}}(\mathrm{X}_1 - \mathbf{1}\bar{\mathbf{x}}_1^{\mathrm{T}}) \tag{9.18}$$

Dividing by $(n-1)$, Equation (9.18) in terms of the partitions of S becomes:

$$S_{22}\hat{B} = S_{21} \tag{9.19}$$

The analogy with the univariate case follows through with sums of squares replaced by SSP matrices, again assuming that S_{22}^{-1} exists.

In particular, the residual SSPM matrix $R_{q.(p-q)}$ is given by

$$R_{q.(p-q)} = (n-1)(S_{11} - S_{12}S_{22}^{-1}S_{21}) = (n-1)S_{q.(p-q)} \tag{9.20}$$

which is a $(q \times q)$ matrix on $n - 1 - (p - q) = n - p + q - 1$ d.f.

Note that the estimation of the parameters $\boldsymbol{\mu}_1 = E(\mathbf{X}_1)$ and B still involves $1 + (p - q)$ linear combinations of the n q-dimensional random variables $\mathbf{X}_{1r}$; $r = 1, 2, \ldots, n$, whose observations appear in the data matrix X_1. $\hat{\boldsymbol{\mu}}_1$ is given in Equation (9.17), the others may be more easily seen by transposing Equation (9.18) and writing it in the form (since $(X_2 - \mathbf{1}\bar{\mathbf{x}}_2^{\mathrm{T}})^{\mathrm{T}}\mathbf{1} = \mathbf{0}$):

$$\hat{B}^{\mathrm{T}} = X_1^{\mathrm{T}}(X_2 - \mathbf{1}\bar{\mathbf{x}}_2^{\mathrm{T}})S_{22}^{-1}/(n-1)$$

which we may write as

$$\hat{B}^{\mathrm{T}} = X_1^{\mathrm{T}}L, \text{ where } L = [\mathbf{l}_{(q+1)}, \mathbf{l}_{(q+2)}, \ldots, \mathbf{l}_p]$$

Hence
$$\hat{B}^{\mathrm{T}} = [X_1^{\mathrm{T}}\mathbf{l}_{(q+1)}, X_1^{\mathrm{T}}\mathbf{l}_{(q+2)}, \ldots, X_1^{\mathrm{T}}\mathbf{l}_p]$$

where $X_1^{\mathrm{T}}\mathbf{l}_{(q+j)} = \sum_{r=1}^{n} l_{r,q+j}x_{1r}, \ j = 1,2,\ldots,p-q.$

As before, the residual SSPM provides the estimate of $\Sigma_{q.(p-q)}$ given by

$$\hat{\Sigma}_{q.(p-q)} = R_{q.(p-q)}/(n-p+q-1) \tag{9.21}$$

and

$$\text{SSPM (Regression)} = (n-1)S_{12}S_{22}^{-1}S_{21} \tag{9.22}$$

on $(p-q)$ d.f.

However, we are unable to write a simple expression for $\mathrm{Var}(\hat{B})$ analagous to Equation (9.14) since $\hat{B} = [\hat{\boldsymbol{\beta}}_1, \hat{\boldsymbol{\beta}}_2, \ldots, \hat{\boldsymbol{\beta}}_q]$ and $\mathrm{Var}(\hat{\boldsymbol{\beta}}_j)$, $j = 1, 2, \ldots, q$, are themselves matrices. From Equation (9.18) we see that

$$\begin{aligned}(n-1).\hat{\boldsymbol{\beta}}_j &= S_{22}^{-1}(X_2 - \mathbf{1}\bar{\mathbf{x}}_2^{\mathrm{T}})^{\mathrm{T}}(\mathbf{y}_j - \mathbf{1}\bar{y}_{.j}), \quad j = 1, 2, \ldots, q\\ &= S_{22}^{-1}(X_2 - \mathbf{1}\bar{\mathbf{x}}_2^{\mathrm{T}})^{\mathrm{T}}\mathbf{y}_j \quad \text{since } (X_2 - \mathbf{1}\bar{\mathbf{x}}_2^{\mathrm{T}})^{\mathrm{T}}\mathbf{1} = \mathbf{0}\end{aligned}$$

Hence, using Equation (9.16),

$$\begin{aligned}\mathrm{Cov}(\hat{\boldsymbol{\beta}}_j, \hat{\boldsymbol{\beta}}_k) &= S_{22}^{-1}(X_2 - \mathbf{1}\bar{\mathbf{x}}_2^{\mathrm{T}})^{\mathrm{T}}\sigma_{jk.(p-q)}I_n(X_2 - \mathbf{1}\bar{\mathbf{x}}_2^{\mathrm{T}})S_{22}^{-1}/(n-1)^2\\ &= \sigma_{jk.(p-q)}S_{22}^{-1}/(n-1)\end{aligned} \tag{9.23}$$

We shall not derive the above results. A derivation based on least squares may be found in Rao (1973, Section 8c) and the maximum-likelihood theory is given in Anderson (1958, Chapter 8). The least-squares estimates of the regression coefficients are exactly the same as those obtained by considering the separate regression of each component of the dependent variable on the regressors, and this fact enables a conceptual understanding to be based on univariate experience. The multivariate aspects appear in the variance structure and in Equation (9.23) in a fairly obvious manner.

We shall now extend the analogy to univariate regression by summarizing the above results in a MANOVA table (see Table 9.1).

The SSPM are Wishart matrices. $R_{q.(p-q)} \sim W_q(n-p+q-1, \Sigma_{q.(p-q)})$, a

Table 9.1

Source	d.f.	SSPM	E(SSPM)
Regression	$p-q$	$(n-1)S_{12}S_{22}^{-1}S_{21}$	$(p-q)\Sigma_{q.(p-q)} + (n-1)B^{\mathrm{T}}S_{22}B$
Residual	$n-p+q-1$	$R_{q.(p-q)}$	$(n-p+q-1)\Sigma_{q.(p-q)}$
Total	$n-1$	$(n-1)S_{11}$	

central distribution, and SSPM(Regression) $\sim W_q(p-q, \Sigma_{q.(p-q)}; GB)$, a non-central distribution where $G = (X_2 - \mathbf{1}\bar{\mathbf{x}}_2^T)$ and $G^T G = (n-1)S_{22}$ as before. These results may be established by a suitable choice of an orthogonal matrix leading to a partition of $(n-1)S_{11}$, the matrix being exactly the same as one suitable for the partitioning of χ^2 in the univariate case. When $q = 1$, Table 9.1 becomes the ANOVA table for a univariate multiple regression of Y_1 on $Z_2, Z_3, \ldots, Z_p$.

A test of the null hypothesis $H_0: B = 0$ (or equivalently $\Sigma_{12} = 0$) is provided by the Wilks Λ-test with

$$\Lambda_{q.(p-q),(p-q),n-p+q-1} = \frac{|R_{q.(p-q)}|}{|\text{SSPM(Regression)} + R_{q.(p-q)}|}$$

$$= \frac{|S_{11} - S_{12}S_{22}^{-1}S_{21}|}{|S_{11}|}$$

$$= \frac{|S|}{|S_{11}||S_{22}|} \quad \text{by Equation (1.4)} \qquad (9.24)$$

and we see that the statistic may be easily calculated without calculating the regression estimates. This result is a special case of Theorem 9.2.1 in Anderson (1958), and is just the likelihood-ratio statistic for testing the independence of $\mathbf{X}_1$ and $\mathbf{X}_2$.

Example 9.1. The data of Table 7.2 refer to girls of approximately two years of age. Suppose their ages are known (to the nearest month) to be as follows:

Girl number	1	2	3	4	5	6	7	8	9
Age (months)	21	27	27	22	26	26	23	22	22

Let us reorder the variables so that our four-dimensional measurement is given by $\mathbf{X}^T = [\mathbf{X}_1^T, \mathbf{X}_2^T] = $ [Chest circumference, MUAC¦Height, Age].

After the necessary additional calculations we have:

$$\bar{\mathbf{x}}^T = [\bar{\mathbf{x}}_1^T, \bar{\mathbf{x}}_2^T] = [58.4, 13.5 \,\vdots\, 76.0, 24.0]$$

and

$$(n-1)S = 8S = \left[\begin{array}{cc:cc} 15.76 & 11.65 & 45.10 & 16.30 \\ 11.65 & 14.50 & 34.50 & 19.50 \\ \hdashline 45.10 & 34.50 & 196.00 & 20.00 \\ 16.30 & 19.50 & 20.00 & 48.00 \end{array}\right]$$

Consider the regression of $\mathbf{X}_1$ on $\mathbf{X}_2$.

The test for significance of regression is given by

$$\Lambda_{2.2,2,8-2} = \frac{|S|}{|S_{11}||S_{22}|} = \frac{|(n-1)S|}{|(n-1)S_{11}||(n-1)S_{22}|}$$

$$= \frac{61{,}063.05}{92.7975 \times 9008}$$

i.e., $\Lambda_{2.2,2,6} = 0.0730$.

We shall use the distributional approximations in Equations (8.18)–(8.21) to obtain significance levels, but note that in those formulae *we must replace p by q* since we are now dealing with the q-dimensional random variable $\mathbf{X}_1|(\mathbf{X}_2 = \mathbf{x}_2)$.

Using Equation (8.18): $-\left(6 - \frac{2-2+1}{2}\right)\ln \Lambda = 14.39;\ qh = 4;$

$$\chi^2(4) = 13.28\ (P = .006)$$

Or, using Equation (8.20): $\frac{1-\Lambda^{1/2}}{\Lambda^{1/2}}\frac{6-2+1}{2} = 6.75;$

$$F_{.01}(4{,}10) = 5.99\ (P = .007)$$

Both approximations indicate significance at the one per cent level, so providing strong evidence to reject the null hypothesis that $B = 0$. Proceeding with the estimation, we obtain

$$\hat{B} = S_{22}^{-1}S_{21}$$

$$= \frac{1}{9008}\begin{bmatrix} 48 & -20 \\ -20 & 196 \end{bmatrix}\begin{bmatrix} 45.10 & 34.50 \\ 16.30 & 19.50 \end{bmatrix} = \begin{bmatrix} 0.2041 & 0.1405 \\ 0.2545 & 0.3477 \end{bmatrix}$$

(When $q = p - q$ as in this example, take care to use S_{21} and not S_{12} in error.)

Hence the fitted regressions are

$$\hat{x}_{1r} - 58.4 = 0.2041(x_{3r} - 76.0) + 0.2545(x_{4r} - 24.0)$$

and $$\hat{x}_{2r} - 13.5 = 0.1405(x_{3r} - 76.0) + 0.3477(x_{4r} - 24.0)$$

The residual matrix $R_{2.2} = (n-1)[S_{11} - S_{12}S_{22}^{-1}S_{21}]$

$$= (n-1)S_{11} - [(n-1)S_{12}][(n-1)S_{22}]^{-1} \times [(n-1)S_{21}]$$

$$= \begin{bmatrix} 15.76 & 11.65 \\ 11.65 & 14.50 \end{bmatrix} - \begin{bmatrix} 13.355 & 12.006 \\ 12.006 & 11.629 \end{bmatrix}$$

$$= \begin{bmatrix} 2.405 & -0.356 \\ -0.356 & 2.871 \end{bmatrix}$$

on $8 - 2 = 6$ d.f.

Hence

$$\hat{\Sigma}_{2.2} = R_{2.2}/6 = \begin{bmatrix} 0.40 & -0.09 \\ -0.09 & 0.48 \end{bmatrix}$$

estimates $\mathrm{Var}(\mathbf{X}_1|\mathbf{x}_2)$, the conditional covariance matrix.
We can compare this with

$$\hat{\Sigma}_{11} = \begin{bmatrix} 1.97 & 1.46 \\ 1.46 & 1.81 \end{bmatrix}$$

which estimates $\mathrm{Var}(\mathbf{X}_1)$, the unconditional covariance matrix.

The reduction in variance is considerable and the correlation between the conditional variates is negligible. We can examine the individual regressions of chest circumference and MUAC and calculate the univariate variance ratio statistics. For chest circumference, SS(Residual) = 2.405 on 6 d.f. and SS(Regression) = 13.355 on 2 d.f., giving a variance ratio of 16.66 with $F_{.01}(2,6) = 10.9$. Similarly for MUAC the variance ratio is 12.15. □

9.3 Canonical correlation

Once the non-independence of $\mathbf{X}_1$ and $\mathbf{X}_2$ is established, a detailed examination is required to identify the more important relationships between the two sets of components. In an application where p and q are small (as in Example 9.1), the obvious follow-up procedure is to examine the regressions of the components of $\mathbf{X}_1$. However, when p and q are large, too much detail is available and the picture is usually too complicated to summarize easily. Instead, we can find that linear compound of $\mathbf{X}_1$ which is most dependent, in a linear-regression sense, on the values of $\mathbf{X}_2$. Specifically, if $U = \mathbf{a}^{\mathrm{T}}\mathbf{X}_1$, we wish to choose $\mathbf{a}$ so that in a regression of U on $X_{q+1}, X_{q+2}, \ldots, X_p$, the variance ratio for testing the significance of the regression is a maximum.

This is nothing other than finding the first canonical variate as in Section 8.6 where the design matrix is in terms of the (given) values of X_2. Following the procedure of Section 8.6, we consider Equation (8.28) $(H - \theta R)\mathbf{a} = \mathbf{0}$. In this application this becomes:

$$(n-1)[S_{12}S_{22}^{-1}S_{21} - \theta(S_{11} - S_{12}S_{22}^{-1}S_{21})]\mathbf{a} = \mathbf{0}$$

which may be simplified to

$$[S_{12}S_{22}^{-1}S_{21} - \nu S_{11}]\mathbf{a} = \mathbf{0} \quad \text{where } \nu = \theta/(1+\theta) \tag{9.25}$$

To obtain a unique solution of Equation (9.25) we must add the constraint

$$\mathbf{a}^{\mathrm{T}}S_{11}\mathbf{a} = 1 \tag{9.26}$$

This is just Equation (8.25) modified since we are working with variance estimators, and it implies that, for the n values of U corresponding to the n

observations on $\mathbf{X}_1$, $\hat{\text{Var}}(U)=1$. The n values of U are given by $\mathbf{u}=X_1\mathbf{a}$ in terms of the data matrix.

From Equation (9.7), replacing $\mathbf{y}_1$ by $\mathbf{u}=\mathrm{X}_1\mathbf{a}$, we obtain

$$(n-1)S_{22}\hat{\boldsymbol{\beta}}=(X_2-\mathbf{1}\bar{\mathbf{x}}_2^{\mathrm{T}})^{\mathrm{T}}(\mathbf{u}-\mathbf{1}\bar{u}_{.})=(X_2-\mathbf{1}\bar{\mathbf{x}}_2^{\mathrm{T}})^{\mathrm{T}}(X_1-\mathbf{1}\bar{\mathbf{x}}_1^{\mathrm{T}})\mathbf{a}=(n-1)S_{21}\mathbf{a}$$

so that the vector of fitted values is given by

$$\hat{\mathbf{u}}=\hat{E}(\mathbf{U}|X_2)=\mathbf{1}\bar{u}_{.}+(X_2-\mathbf{1}\bar{\mathbf{x}}_2^{\mathrm{T}})S_{22}^{-1}S_{21}\mathbf{a} \tag{9.27}$$

It is convenient to consider the above regression as a regression of U on a linear compound of $\mathbf{X}_2$. We will denote this linear compound by $V=\mathbf{b}^{\mathrm{T}}\mathbf{X}_2$, where $\mathbf{b}$ must be proportional to $\hat{\boldsymbol{\beta}}$, and scale $\mathbf{b}$ so that

$$\mathbf{b}^{\mathrm{T}}S_{22}\mathbf{b}=1 \tag{9.28}$$

If $\mathbf{b}=k\hat{\boldsymbol{\beta}}=kS_{22}^{-1}S_{21}\mathbf{a}$, then using Equations (9.25), (9.26) and (9.28)

$$\mathbf{b}^{\mathrm{T}}S_{22}\mathbf{b}=k^2\mathbf{a}^{\mathrm{T}}S_{12}S_{22}^{-1}S_{21}\mathbf{a}=k^2\nu\mathbf{a}^{\mathrm{T}}S_{11}\mathbf{a}=k^2\nu=1$$

so that $k=\nu^{-1/2}$ and

$$\mathbf{b}=\nu^{-1/2}S_{22}^{-1}S_{21}\mathbf{a}=\nu^{-1/2}\hat{B}\mathbf{a} \tag{9.29}$$

Hence we define

$$V=\nu^{-1/2}\mathbf{a}^{\mathrm{T}}S_{12}S_{22}^{-1}\mathbf{X}_2$$

taking the n values $\mathbf{v}=\nu^{-1/2}X_2S_{22}^{-1}S_{21}\mathbf{a}$. Consequently we may write the fitted regression Equation (9.27), as

$$\hat{u}_r=\hat{E}(U_r|X_2)=\bar{u}_{.}+\nu^{1/2}(v_r-\bar{v}_{.}) \qquad r=1,2,\ldots,n \tag{9.30}$$

Now $S_{12}\mathbf{b}=\nu^{-1/2}S_{12}S_{22}^{-1}S_{21}\mathbf{a}=\nu^{1/2}S_{11}\mathbf{a}$, by Equation (9.25), so that

$$\mathbf{a}=\nu^{-1/2}S_{11}^{-1}S_{12}\mathbf{b} \tag{9.31}$$

Compare this with Equation (9.29).

Also

$$S_{22}\mathbf{b}=\nu^{-1/2}S_{21}\mathbf{a}=\nu^{-1}S_{21}S_{11}^{-1}S_{12}\mathbf{b}$$

so that

$$[S_{21}S_{11}^{-1}S_{12}-\nu S_{22}]\mathbf{b}=\mathbf{0} \tag{9.32}$$

Compare this with Equation (9.25).

The symmetry of these relationships is a reflection of the existence of two possible regressions associated with the bivariate distribution of U and V, namely the regression of U on V and the regression of V on U.

The relationship between U and V is often best expressed in terms of the correlation coefficient. To examine the relationship in terms of correlation, note first that

$$\hat{\text{Cov}}(U,V)=\hat{\text{Cov}}(\mathbf{a}^{\mathrm{T}}\mathbf{X}_1,\mathbf{b}^{\mathrm{T}}\mathbf{X}_2)=\mathbf{a}^{\mathrm{T}}S_{12}\mathbf{b}$$

so that the estimated correlation coefficient is

$$
\begin{aligned}
r(U, V) &= \mathbf{a}^{\mathrm{T}}S_{12}\mathbf{b}/\sqrt{(\mathbf{a}^{\mathrm{T}}S_{11}\mathbf{a} \times \mathbf{b}^{\mathrm{T}}S_{22}\mathbf{b})} \\
&= \mathbf{a}^{\mathrm{T}}S_{12}\mathbf{b} \qquad \text{by Equations (9.26) and (9.28)} \\
&= v^{-1/2}\mathbf{a}^{\mathrm{T}}S_{12}S_{22}^{-1}S_{21}\mathbf{a} \qquad \text{by Equation (9.29)} \\
&= v^{1/2} \qquad \text{by Equations (9.25) and (9.26)}
\end{aligned} \tag{9.33}
$$

U and V have maximum correlation when v is the greatest root of Equations (9.25) or (9.32). These first canonical variates will be denoted by $U_1 = \mathbf{a}_1^{\mathrm{T}}\mathbf{X}_1$ and $V_1 = \mathbf{b}_1^{\mathrm{T}}\mathbf{X}_2$ where $\mathbf{a}_1$ and $\mathbf{b}_1$ are the solutions of Equations (9.25) and (9.32) when $v = v_1$. Corresponding to the other non-zero roots of Equation (9.25) we can find pairs of correlated variates (U_j, V_j) with correlation coefficient $v_j^{1/2}$ and, as with the canonical variates of Section 8.6, the U_j are uncorrelated with each other, as are the V_j.

The canonical variates U and V are frequently defined as arising from the problem of maximizing $r(U, V)$ for all $\mathbf{a}$ and $\mathbf{b}$, which is equivalent to the problem of maximizing $\mathbf{a}^{\mathrm{T}}S_{12}\mathbf{b}$, subject to the constraints $\mathbf{a}^{\mathrm{T}}S_{11}\mathbf{a} = \mathbf{b}^{\mathrm{T}}S_{22}\mathbf{b} = 1$. Differentiating the Lagrangian

$$L(\mathbf{a}, \mathbf{b}) = \mathbf{a}^{\mathrm{T}}S_{12}\mathbf{b} + \alpha(1 - \mathbf{a}^{\mathrm{T}}S_{11}\mathbf{a}) + \beta(1 - \mathbf{b}^{\mathrm{T}}S_{22}\mathbf{b})$$

with respect to $\mathbf{a}$ and $\mathbf{b}$ leads to the equations (see Anderson, 1958)

$$
\begin{aligned}
-\rho S_{11}\mathbf{a} + S_{12}\mathbf{b} &= \mathbf{0} \\
S_{12}\mathbf{a} - \rho S_{22}\mathbf{b} &= \mathbf{0}
\end{aligned} \tag{9.34}
$$

where $\rho = \mathbf{a}^{\mathrm{T}}S_{12}\mathbf{b} = r(U, V)$.

Elimination of $\mathbf{b}$ and $\mathbf{a}$ in turn leads to Equations (9.25) and (9.32) with $\rho^2 = v$. ρ is the *canonical correlation* (see Equation (9.33) above).

Note that S_{11} and S_{22} are assumed to be of full rank in the regression appproach and that in practice the number of non-zero roots will be the minimum of q and $(p - q)$. However, the assumption of full rank is not necessary for the derivation of the canonical variates.

To understand the general case it is helpful to think of the data matrix with the variates measured about their means i.e., replace X by $(X - \mathbf{1}\bar{\mathbf{x}}^{\mathrm{T}})$ as a set of p n-dimensional vectors. These vectors lie in a vector space of dimension $(n - 1)$, say $\mathscr{W}$. The $(p - q)$ vectors in X_2 (representing the values of the covariates) define a subspace $\mathscr{V} \subset \mathscr{W}$ and S_{22} is of full rank, $(p - q)$, when they are linearly independent. Similarly, the q vectors in X_1 (representing the observations on the response) define a subspace $\mathscr{U} \subset \mathscr{W}$.

Now, any vector $\mathbf{w} \in \mathscr{W}$ can be expressed as the sum of two components, one lying in $\mathscr{V}$, the other orthogonal to $\mathscr{V}$. The component in $\mathscr{V}$ is simply the regression of $\mathbf{w}$ on the covariates or, in geometrical terms, the projection of $\mathbf{w}$ on $\mathscr{V}$. The projection is unique but it can be expressed uniquely in terms of X_2 only when S_{22} is of full rank. The component orthogonal to $\mathscr{V}$ is just the vector of residuals about the regression.

Suppose that $\mathbf{u}$ belongs to $\mathscr{U}$ and is of the form $\mathbf{u} = X_1\mathbf{a}$ where $\mathbf{a}$ is chosen so that $\mathbf{a}^T S_{11} \mathbf{a} = 1$. The first canonical variate, $\mathbf{u}_1 = X_1 \mathbf{a}_1$ is that vector in $\mathscr{U}$ which has the greatest component in $\mathscr{V}$. The projection of $\mathbf{u}_1$ on $\mathscr{V}$ is proportional to the corresponding $\mathbf{v}_1$. The angle between $\mathbf{u}_1$ in $\mathscr{U}$ and $\mathbf{v}_1$ in $\mathscr{V}$ is the smallest possible for all $\mathbf{u} \in \mathscr{U}$ and all $\mathbf{v} \in \mathscr{V}$.

In particular, note that if $\mathbf{u}_1$ lies wholly in $\mathscr{V}$, then $\rho = \pm 1$ and conversely if $\rho_1 = \pm 1$, $\mathbf{u}_1$ can be expressed exactly as a linear combination of the covariate vectors. If the subspaces $\mathscr{U}$ and $\mathscr{V}$ are orthogonal, then any $\mathbf{u}$ is uncorrelated with any $\mathbf{v}$ and $\mathbf{X}_1$ is uncorrelated with $\mathbf{X}_2$. In this case all the ρ_j are zero.

It can be shown that the number of non-zero canonical correlations is given by rank (Σ_{12}) (see Rao, 1973, Section 8f.1). In analysing data with Σ replaced by its estimate S, a test of hypothesis is required to determine the number of non-zero canonical correlations. Bartlett's test of dimensionality (see Equation (8.23)) can be used with λ_j replaced by $(1 - v_j) = (1 - \rho_j^2)$ and with p replaced by q since $\mathbf{X}_1 | \mathbf{X}_2$ is q-dimensional. The d.f. for residual is $r = n - p + q - 1$ and for hypothesis is $h = p - q$ (see Table 9.1 above), so that

$$-\left(n - \frac{p+3}{2}\right) \sum_{j=t+1}^{q} \ln(1 - v_j) \sim \chi^2\{(q-t)(p-q-t)\} \tag{9.35}$$

under the null hypothesis that only t of the canonical correlation are non-zero.

Example 9.2. Using the data of Example 9.1, let us calculate the canonical correlations between $\mathbf{X}_1$ and $\mathbf{X}_2$ where $\mathbf{X}_1^T = $ [Chest circumference, MUAC] and $\mathbf{X}_2^T = $ [Height, Age].

The roots v_1 and v_2 may be determined from Equation (9.25) or from Equation (9.32) and we would naturally choose to use that equation involving matrices of lower order, using (9.25) if $q < (p - q)$ and (9.32) if $q > (p - q)$. Here $q = p - q = 2$, so the choice is immaterial.

Now

$$(n-1)S_{11} = \begin{bmatrix} 15.76 & 11.65 \\ 11.65 & 14.50 \end{bmatrix}$$

and we have previously calculated (Example 9.1)

$$(n-1)S_{12}S_{22}^{-1}S_{21} = \begin{bmatrix} 13.3551 & 12.0058 \\ 12.0058 & 11.6287 \end{bmatrix}$$

Hence, using Equation (9.25), v_1 and v_2 are the roots of:

$$\begin{vmatrix} (13.3551 - 15.76v) & (12.0058 - 11.65v) \\ (12.0058 - 11.65v) & (11.6287 - 14.50v) \end{vmatrix} = 0$$

and we obtain $v_1 = 0.9159$, $v_2 = 0.1313$.

We note that v_1 is quite large, the canonical correlation being $v_1^{1/2} = 0.96$ whereas $v_2^{1/2} = 0.36$. It seems that only the first canonical variates will be of practical importance in examining the dependence of $\mathbf{X}_1$ on $\mathbf{X}_2$. Note that $(1 - \nu_1)(1 - \nu_2) = 0.0730 = \Lambda_{2.2,2,6}$. as calculated in Example 9.1.

Applying the test in Equation (9.35) we obtain for $t = 1$:

$$-\left(9 - \frac{4+3}{2}\right)\ln(1 - 0.1313) = 0.77$$

$$(q-t)(p-q-t) = 1 \qquad \chi^2_{.05}(1) = 3.84$$

The result is not significant and it would not be unreasonable to assume $\nu_2 = \rho_2^2 = 0$. When $t = 0$ we obtain the overall test and in Example 9.1 we found that this was significant at the 1 per cent level. We can feel sure that $\nu_1 = \rho_1^2$ is estimating a real correlation.

Now let us find the canonical variates U_1 and V_1. Putting $\nu = 0.9159$ in Equation (9.25) and writing $\mathbf{a}_1^{\mathrm{T}} = k[a, 1]$ we obtain

$$-1.0796a + 1.3355 = 0$$
$$1.3355a - 1.6520 = 0$$

and hence $a = 1.2370$.

Substituting in Equation (9.26),

$$k^2[a, 1]\begin{bmatrix} 15.76 & 11.65 \\ 11.65 & 14.50 \end{bmatrix}\begin{bmatrix} a \\ 1 \end{bmatrix}\Big/ 8 = 1$$

giving $k = 0.3444$, $ka = 0.4261$, so that

$$U_1 = 0.4261\text{ (Chest circumference)} + 0.3444\text{(MUAC)}.$$

$\mathbf{b}_1$ is obtained from Equation (9.29) and we find

$$V_1 = 0.1415\text{(Height)} + 0.2384\text{(Age)}$$

$$\bar{u} = [0.4261 \quad 0.3444]\begin{bmatrix} 58.4 \\ 13.5 \end{bmatrix} = 29.53$$

and similarly, $\bar{v} = 16.47$.

Hence the regression of U on V is, by Equation (9.30),

$$\hat{u}_r - 29.53 = 0.96(v_r - 16.47)$$

and the regression of V on U is

$$\hat{v}_r - 16.47 = 0.96(u_r - 29.53)$$ □

9.4 The multivariate analysis of covariance

9.4.1 *The special case: Univariate analysis of covariance*

Let us begin this section by reviewing the methodology of analysis of covariance for a univariate response and a single covariate. We shall do this by

Table 9.2

Source	d.f.	Corrected sums of squares and products		
		S(zz)	S(zx)	S(xx)
Temperature (T)	2	4.8161	.28437	.019633
Sex (S)	1	.6422	−.19644	.060089
$T \times S$	2	.2755	.03818	.006078
Residual	12	19.3264	−.19063	.039200
Total	17	25.0602	−.06453	.125000

considering an example for which we shall use part of the data of Example 8.2. This example refers to a two-factor fully randomized experiment on mice. Suppose we regard tumour weight (X_3) as the response and denote it by X, and the initial weight of the animal (X_1) as the covariate and denote it by z. We shall, for the moment, ignore the data for the final weight (X_2).

The calculation for an analysis of covariance (ANOCOVA) may be abstracted from Table 8.7 and is set out in Table 9.2.

The model to be fitted is

$$\Omega : X_{ijk} = \mu + \alpha_i + \beta_j + \theta_{ij} + \gamma z_{ijk} + e_{ijk} \qquad (9.36)$$
$$i = 1,2,3; \quad j = 1,2; \quad k = 1,2,3.$$

α_i is the ith temperature effect, β_j the jth sex effect and θ_{ij} the (i, j)th interaction. γ is the regression coefficient for the regression of response on initial weight and is assumed to be the same for all treatments. The e_{ijk} are independently distributed as N(0, $\sigma^2_{X|z}$).

$X_{ijk} - \hat{\gamma}(z_{ijk} - \bar{z}_{...})$ is the response adjusted for initial weight, the adjustment resulting in data from mice considered to have had the same initial weight $\bar{z}_{...}$. The choice of $\bar{z}_{...}$ is conventional and it could be replaced by any constant value.

The least-squares estimate for γ is calculated using the residuals after fitting the other parameters and so is obtained from the residual line in Table 9.2. Here

$$\hat{\gamma} = [\mathrm{S}(zx)/\mathrm{S}(zz)]_{(\mathrm{Residual})} = -0.19063/19.3264 = -0.0099 \qquad (9.37)$$

and this is used in the calculation of adjusted treatment means. For example, indicating adjusted means by an asterisk,

$$\bar{x}^*_{ij.} = \bar{x}_{ij.} - \hat{\gamma}(\bar{z}_{ij.} - \bar{z}_{...})$$

The variance about the fitted model, denoted by $\sigma^2_{X|z}$ is estimated by the residual mean square of the adjusted response. In this example, the residual sum of squares is

$$R(\Omega) = [\mathrm{S}(xx) - \{[\mathrm{S}(zx)]^2/\mathrm{S}(zz)\}]_{(\mathrm{Residual})} \qquad (9.38)$$

Note that this is the usual linear regression residual calculated using the residual line in Table 9.2. The d.f. associated with $R(\Omega)$ is given by [d.f. (residual) – the number of covariates]. Here $R(\Omega)$ is based on $12 - 1 = 11$ d.f.

We can examine the fitted model to see whether or not the covariate has made a useful contribution to the analysis by testing the null hypothesis $H_0: \gamma = 0$. The test is displayed in an ANOVA table, as shown in Table 9.3.

Table 9.3 Test for significance of dependence on covariate

Source	d.f.	SS	MS	V.R.
Regression	1	$\{[S(zx)]^2/[S(zz)]\} = 0.001880$	0.001880	< 1
Residual$\mid\Omega$	11	$R(\Omega) = 0.037320$	0.003393	
Residual$\mid\Omega_0$	12	$S(xx) = 0.039200$		

We see that there is no evidence that the tumour weight depends on the initial weight. In practice we would now abandon the covariate and continue with an ANOVA of the tumour weights. However, for the purpose of illustrating the methodology we shall ignore the results of this test and continue with the analysis of covariance.

Suppose we wish to test the null hypothesis $H_0: \theta_{ij} = 0$ for all i and j. Following the basic least-squares approach, we shall fit the model incorporating the null hypothesis:

$$\Omega_0 : X_{ijk} = \mu + \alpha_i + \beta_j + \gamma z_{ijk} + e_{ijk}$$

The equivalent of Table 9.2 will contain the same lines for 'Temperature', 'Sex' and 'Total', but since there is no interaction line the residual line for Ω_0 is obtained by summing the lines for '$T \times S$' and 'Residual' in Table 9.2. This is presented in Table 9.4 together with the adjusted residuals calculated using Equation (9.38).

In a univariate analysis the test statistic is

$$\text{V.R.} = \frac{[R(\Omega_0) - R(\Omega)]/(r_0 - r)}{R(\Omega)/r}$$

where r and r_0 denote the d.f. for the residuals.

Table 9.4 Test for significance of the interaction

Source	d.f.	Corrected sums of squares and products			Adjusted residual	d.f.	MS	V.R.
		S(zz)	S(zx)	S(xx)				
$T \times S$	2	.2755	.03818	.006078	$R(\Omega_0) - R(\Omega) = .006772$	2	.003386	1.00
Residual$\mid\Omega$	12	19.3264	–.19063	.039200	$R(\Omega) = .037320$	11	.003393	
Residual$\mid\Omega_0$	14	19.6019	–.15245	.045278	$R(\Omega_0) = .044092$	13		

The statistic could be compared with the tabled values of $F(2, 11)$ but, here, is obviously not significant.

The analysis would now continue with similar tests on the main effects.

9.4.2 *The multivariate case: An example*

Now let us turn to the multivariate case and analyse the full data of Example 8.2. Again initial weight (X_1) will be regarded as a covariate again denoted by z, but this time we shall consider the bivariate response $\mathbf{X}$ where $\mathbf{X}^T = [X_2 \quad X_3]$. The model Ω becomes:

$$\Omega : \mathbf{X}_{ijk} = \boldsymbol{\mu} + \boldsymbol{\alpha}_i + \boldsymbol{\beta}_j + \boldsymbol{\theta}_{ij} + \boldsymbol{\gamma} z_{ijk} + \mathbf{e}_{ijk}$$

The parameters are vectors of dimension $q = 2$, γ is the $(q \times 1)$ vector of the regression coefficients of the regressions of the $q = 2$ components of the response on the single covariate (here $p - q = 3 - 2 = 1$). In general, where there are $p - q$ covariates, the regression coefficients would be the elements of a $(q \times (p - q))$ matrix C, say.

The equivalent of Table 9.2 giving the corrected sums of squares and products is, in fact, the MANOVA table, namely Table 8.7, and γ is estimated from the residual matrix.

Write $$R = \begin{bmatrix} R_{zz} & R_{zx} \\ R_{xz} & R_{xx} \end{bmatrix} \quad \text{on } f = 12\text{d.f.}$$

where R_{zz} is (1×1), R_{xx} is (2×2) and $R_{zx} = R_{xz}^T$ is (1×2).

Then $$\hat{\gamma}^T = R_{zz}^{-1} R_{zx}. \tag{9.39}$$

Compare this with Equation (9.37).

The estimated adjusted response is $\mathbf{X}_{ijk} - \hat{\gamma}(z_{ijk} - \bar{z}_{...})$.

The adjusted residual sum of squares is

$$R(\Omega) = R_{2.1} = R_{xx} - R_{xz} R_{zz}^{-1} R_{zx} \tag{9.40}$$

on $f - (p - q) = 12 - 1 = 11$ d.f.

Using the results of Section 9.1 we can test the usefulness of the covariate as shown in Table 9.5, which is similar to Table 9.1; but note that here R refers to

Table 9.5 Test for significance of dependence on covariate

Source	d.f.	SSPM
Regression	$p - q = 1$	$R_{xz} R_{zz}^{-1} R_{zx} = \begin{bmatrix} 2.5422 & -.06914 \\ -.06914 & .001880 \end{bmatrix}$
Residual	$f - (p - q) = 11$	$R(\Omega) = \begin{bmatrix} 24.1566 & .27761 \\ .27761 & .037320 \end{bmatrix}$
Total	$f = 12$	$R_{xx} = \begin{bmatrix} 26.6988 & .20847 \\ .20847 & .039200 \end{bmatrix}$

corrected sums of squares and products whereas in Table 9.1 the results are partially expressed in terms of S, the estimate of Σ.

Calculating directly, we have

$$\Lambda_{2.1,1,11} = \frac{|R_{2.1}|}{|R_{xx}|} = 0.8219$$

Alternatively, using Equation (9.24) with $f = 12$, $h = 1$,

$$\Lambda_{2.1,1.11} = \frac{|R|}{|R_{zz}||R_{xx}|} = 0.8219$$

There is thus no need to write out Table 9.5. Using Equation (8.18),

$$-\left(r - \frac{q-h+1}{2}\right)\ln \Lambda = -\left(11 - \frac{2-1+1}{2}\right)\ln \Lambda = 1.96$$

Since $\chi^2_{.05}(qh) = \chi^2_{.05}(2) = 5.99$, the result is not significant ($P = 0.38$). Using Equation (8.19),

$$\frac{1-\Lambda}{\Lambda}\,\frac{r-q+1}{q} = 1.08 \qquad F_{.05}(2,10) = 4.10 \qquad (P = 0.38)$$

and we obtain the same result.

We shall again ignore this non-significance in order to illustrate the methodology of hypothesis testing in MANOCOVA.

Suppose we wish to test the null hypothesis that there is no temperature by sex interaction. The approach is identical to the univariate case considered earlier.

$R_{2.1}$ is the adjusted residual matrix when model Ω is fitted. We now proceed to fit model Ω_0 which has no interaction parameter. The adjusted residual matrix $R(\Omega_0)$ is calculated using Equation (9.40) applied to the new residual line. The calculations may be set out as in Table 9.6, which should be compared with Table 9.4. In Table 9.6, $(H+R)_{2.1}$ refers to Equation (9.40) applied to the matrix $(H+R)$.

The multivariate test involves the comparison of the two residual matrices using the Λ statistic. We shall write this statistic in the form

$$\Lambda_{q\cdot(p-q),h,f-(p-q)} = \Lambda_{2.1,2,11} \tag{9.41}$$

where f = d.f. (Residual$|\Omega$), and h = d.f. (Hypothesis).

Table 9.6 Test for significance of the interaction

Source	d.f.	SSPM	Adjusted residual	d.f.
$T \times S$	2	H		
Residual$\|\Omega$	12	R	$R(\Omega) = R_{2.1}$	$12-(p-q) = 11$
Residual$\|\Omega_0$	14	$H+R$	$R(\Omega_0) = (H+R)_{2.1}$	$14-(p-q) = 13$

Now by Equation (1.4),

$$|R_{2.1}| = |R_{xx} - R_{xz}R_{zz}^{-1}R_{zx}|$$
$$= |R|/|R_{zz}|$$

with a similar expression for $|(H+R)_{2.1}|$
Hence

$$\Lambda_{2.1,2,11} = \frac{|R_{2.1}|}{|(H+R)_{2.1}|}$$
$$= \frac{|R|}{|H+R|} \bigg/ \frac{|R_{zz}|}{|(H+R)_{zz}|} \tag{9.42}$$

We see that $\Lambda_{2.1,2,11}$ may be obtained as the ratio of the Wilks' statistic calculated using all p variables to the Λ statistic calculated using only the $(p-q)$ covariates. We shall write this relationship briefly as $\Lambda_{q.(p-q),h,f-(p-q)} = \Lambda_p/\Lambda_{(p-q)}$.

We calculated $\Lambda_p = 0.7720$ for the MANOVA in Section 8.3.5 and, since there is only one covariate,

$$\Lambda_{(p-q)} = \frac{19.3264}{0.2755 + 19.3264} = 0.9859$$

This gives $\Lambda_{2.1,2,11} = 0.7720/0.9859 = 0.7830$ and $-[r-(q-h+1)/2]\ln\Lambda = -10.5\ln\Lambda = 2.57$, using Equation (8.18).

The result is not significant when compared with $\chi^2(qh) = \chi^2(4)$. *Note that p is replaced by q in Equation* (8.18) *since we are considering the q-dimensional adjusted response.*

Since the interaction is not significant, we now proceed to test the main effects.

For the effect of temperature we take $H = \text{SSPM(Temperature)}$ on 2 d.f. From Section 8.3, $\Lambda_p = \Lambda_4 = 0.2617$.
From Table 8.7,

$$\Lambda_{(p-q)} = \Lambda_1 = \frac{19.3264}{4.8161 + 19.3264} = 0.8005.$$

Hence $\Lambda_{2.1,2,11} = 0.2617/0.8005 = 0.3269$.

Using Equation (8.18), $-10.5\ln\Lambda = 11.74$, $\chi^2_{.025}(4) = 11.14$, and the result is significant at the 2.5 per cent level. We note that this is much the same as the result of the test in Section 8.3. This is not surprising since the regression on the covariate was found to be not significant.

A similar test may be carried out for the sex main effect. This will be left as an exercise [$\Lambda_{2.1,1,11} = 0.3485$].

9.4.3 *The multivariate case: General results*

We have introduced MANOCOVA through an example. Let us now summarize the procedure for any designed experiment. The model to be fitted may be written in terms of the response variate, $\mathbf{X}_1$, as

$$\Omega : \mathbf{X}_{1r} = \Theta^{\mathrm{T}}\mathbf{g}_r + C\mathbf{x}_{2r} + \mathbf{e}_r \qquad r = 1, 2, \ldots, n \tag{9.43}$$

$\mathbf{g}_r$ of order $(k \times 1)$ is specified by the design of the experiment (common to all components of the response). Θ^{T} is the $(q \times k)$ matrix of parameters, the ith row giving the parameters for the ith component of the response. $C = ((\gamma_{i,q+j}))$ is the $(q \times (p-q))$ matrix of regression coefficients and the $\mathbf{e}_r$ are independently $\mathrm{N}_q(0, \Sigma_{1.2})$ where $\Sigma_{1.2}$ is the variance of $\mathbf{X}_1$ adjusted for $\mathbf{x}_2$.

The data matrix $X = [X_1, X_2]$ holds the observed values of $\mathbf{X}_1$ and the corresponding values of $\mathbf{x}_2$ *as row vectors*. The model may be rewritten in terms of the data matrix as

$$\Omega : E(X_1) = G\Theta + X_2C^{\mathrm{T}} \tag{9.44}$$

where $G^{\mathrm{T}} = [\mathbf{g}_1, \mathbf{g}_2, \ldots, \mathbf{g}_n]$ is of order $(k \times n)$.

The first step in the analysis is to calculate the MANOVA table for all p-variates. The adjusted residual sum of squares is then calculated as

$$R(\Omega) = R_{q.(p-q)} = R_{11} - R_{12}R_{22}^{-1}R_{21} \tag{9.45}$$

the subscripts referring to the partitions of the residual matrix on f d.f. in an obvious way. $R(\Omega)$ has d.f. $r = f - (p - q)$.

The relevance of the covariate may be tested by

$$\Lambda_{q,p-q,f-(p-q)} = \frac{|R|}{|R_{11}||R_{22}|} \tag{9.46}$$

and, if significant, the regression coefficients may be estimated by

$$\hat{C}^{\mathrm{T}} = R_{22}^{-1}R_{21} \tag{9.47}$$

The formal expression for the adjusted data matrix is

$$X_1^* = X_1 - (X_2 - \mathbf{1}\bar{\mathbf{x}}_2^{\mathrm{T}})R_{22}^{-1}R_{21} \tag{9.48}$$

where the response is adjusted so that each experimental unit may be considered to have a value of $\mathbf{x}_2$ given by $\mathbf{x}_2 = \bar{\mathbf{x}}_2$, the mean over all experimental units.

This formal expression may be used to give adjusted means of subsets of experimental units – treatment means, row means, etc.

To test a hypothesis that a particular effect is not present, the SSPM matrix of the corresponding source is denoted by H on h d.f. and the test statistic is

$$\Lambda_{q.(p-q),h,f-(p-q)} = \Lambda_p/\Lambda_{(p-q)} \tag{9.49}$$

where Λ_p and $\Lambda_{(p-q)}$ are the Λ-statistics $(= |\mathrm{R}|/|\mathrm{H}+\mathrm{R}|)$ for the MANOVA tests using respectively all the variables and the covariates only.

The test is carried out using the approximations given in Equations (8.18)–(8.21) with p replaced by q and with $r = f - (p - q)$.

Referring to Table 9.6, we see that that to carry out the test it is not necessary to calculate the difference between the two alternative residuals as is done in the univariate test in Table 9.4. In the univariate case this difference may be called the *adjusted sum of squares* for the hypothesis. Similarly, the matrix obtained by taking the difference between the two alternative residuals in Table 9.6 may be called the *adjusted SSPM* for the hypothesis and denoted by $H_{q.(p-q)}$. It would then be possible to carry out a canonical variates analysis based on the roots of the equation

$$|H_{q.(p-q)} - \theta R_{q.(p-q)}| = 0 \qquad (9.50)$$

9.5 The test for additional information

Most multivariate problems involve large amounts of data and even after a statistical analysis it is difficult to see the overall picture. Many techniques attempt to reduce the dimensionality of the data so that attention can be focused on a subspace which is in some sense best for examining possible patterns. Examples of such techniques are principal components analysis (Chapter 4), canonical variates analysis (Chapter 8) and canonical correlation (this chapter). All these techniques esentially produce transformed data by finding optimum linear compounds of the response vector. However, to evaluate these transformed data we need to retain the observed values of *all* the original p variables.

It may happen that in some analysis it is apparent that a particular component of the response is making a negligible contribution. For instance, in Example 8.3 the initial weight seems to be unimportant when the aim is to distinguish between the different treatments. If this is the case, we can reduce the dimensionality of the problem and make further analysis easier if we simply drop the particular component altogether. In addition when planning further experiments or collecting more data, we might decide not to include that component. This last point is especially important if our initial experiment was exploratory in nature, for then the tendency is to include many components just in case they prove useful.

The test for additional information described by Rao (1973, Section 8c.4) is designed to assess the usefulness of particular components of the response when the aim of the analysis is to test hypotheses on the linear model.

Suppose we are testing a null hypothesis on a linear model for which the response is a p-dimensional random variable, $\mathbf{X}$, where $\mathbf{X}^T = [\mathbf{X}_1^T, \mathbf{X}_2^T]$, the null hypothesis being denoted by H_0 with corresponding SSPM H on h d.f. Let us now test whether or not the observations on $\mathbf{X}_2$ make a significant contribution to the test on H_0 *independently of the contribution made by the observations on* $\mathbf{X}_1$.

Since $\mathbf{X}_1$ and $\mathbf{X}_2$ are in general correlated, we can find linear compounds of $\mathbf{X}_1$ to estimate $\mathbf{X}_2$. The information in the *estimated values* is fully taken into account when we use the $\mathbf{X}_1$ data by itself to calculate the test statistic for the test on H_0. Consequently, the *independent* contribution made by $\mathbf{X}_2$ to the test comes from the values of $\mathbf{X}_2$ adjusted for $\mathbf{X}_1$.

Thus the procedure for performing a test of additional information is simply to carry out a MANOCOVA with $\mathbf{X}_1$ *as the covariate*, with test statistic given by

$$\Lambda_{(p-q),q,f-q} = \Lambda_p/\Lambda_q \qquad (9.51)$$

If the resulting test of the hypothesis H_0 gives a non-significant result, we infer that $\mathbf{X}_2$ provides no additional information against H_0 over and above that provided through its correlation with $\mathbf{X}_1$. However, note that $\mathbf{X}_2$ could provide *no additional information* against H_{01}, an hypothesis on an interaction, say, but *could* provide additional information against H_{02}, an hypothesis on a main effect. The test of additional information is specific to the hypothesis H_0.

The reader should consult Rao (1973) for a more formal presentation.

Example 9.3. Let us return to Example 8.2 and look again at the test of the null hypothesis that there are no temperature effects. In Section 8.3, we calculated $\Lambda_{3,2,12} = 0.2617$. In Section 9.4, taking the initial weight as a covariate, we calculated $\Lambda_{2.1,2,11} = 0.3269$. Here, let us ask if tumour weight provides information against the null hypothesis additional to the information supplied by initial and final weight.

We calculate, using Equation (9.51),

$$\Lambda_{1.2,2,10} = \Lambda_3/\Lambda_2$$

$\Lambda_3 = 0.2617$ using all three variates. For Λ_2 we ignore the tumour-weight values and calculate

$$\Lambda_2 = |R|/|H+R| = \begin{vmatrix} 19.3264 & 7.0094 \\ 7.0094 & 26.6988 \end{vmatrix} \Bigg/ \begin{vmatrix} 24.1425 & 16.6734 \\ 16.6734 & 59.2855 \end{vmatrix} = 0.4048$$

Hence $\Lambda_{1.2,2,10} = 0.2617/0.4048 = 0.6465$.

Using Equation (8.18), $-[10-(1-2+1)/2]\ln\Lambda = 4.36$

$\chi^2_{.05}(qh) = \chi^2_{.05}(2) = 5.99$

Although this result is not significant at the 5 per cent level ($P = 0.11$), we would be reluctant to accept that tumour weight provides no additional information. □

Example 9.4. In Example 9.1 we added an additional variate to the data of Table 7.2 by giving the ages of the girls to the nearest month. Let us do the same for Table 7.1 by giving the ages of the boys:

Boy number	1	2	3	4	5	6
Age (months)	26	21	24	22	24	21

After the necessary additional calculations (and re-ordering of the variates so that $\mathbf{X}^{\mathrm{T}} = $[Chest circumference, MUAC, Height, Age]) we obtain, with $n = 6$,

$$\bar{\mathbf{x}}^{\mathrm{T}} = [60.2,\ 14.5,\ 82.0,\ 23.0];$$

$$(n-1)S = \begin{bmatrix} 15.86 & 6.55 & 40.20 & 11.60 \\ 6.55 & 9.50 & 2.50 & 12.50 \\ 40.20 & 2.50 & 158.00 & 6.00 \\ 11.60 & 12.50 & 6.00 & 20.00 \end{bmatrix}$$

Combining these with the corresponding results of Example 9.1 to compare the means for boys and girls, we obtain

$$\bar{\mathbf{d}}^{\mathrm{T}} = \bar{\mathbf{x}}_b^{\mathrm{T}} - \bar{\mathbf{x}}_g^{\mathrm{T}} = [1.8,\ 1.0,\ 6.0,\ -1.0]$$

where the subscript denote 'boys' and 'girls', and

$$(n_b + n_g - 2)S = 13S = \begin{bmatrix} 31.62 & 18.20 & 85.30 & 27.90 \\ 18.20 & 24.00 & 37.00 & 32.00 \\ 85.30 & 37.00 & 354.00 & 26.00 \\ 27.90 & 32.00 & 26.00 & 68.00 \end{bmatrix} \text{ on 13 d.f.}$$

To test $H_0 : \boldsymbol{\mu}_b = \boldsymbol{\mu}_g$ against the alternative that the mean vectors are not equal, the Hotelling T^2-statistic is

$$\mathscr{T}^2 = \frac{9 \times 6}{9 + 6} \bar{\mathbf{d}}^{\mathrm{T}} S^{-1} \bar{\mathbf{d}} \qquad \text{(see Section 7.5)}$$

$$= \frac{54}{15} \bar{\mathbf{d}}^{\mathrm{T}} 13\mathbf{a}^*$$

where $\mathbf{a}^*$ is the solution of $(13\ S)\mathbf{a}^* = \bar{\mathbf{d}}$.

Solving for $\mathbf{a}^*$, we obtain

$$(\mathbf{a}^*)^{\mathrm{T}} = [0.2187,\ 0.1505,\ -0.0397,\ -0.1601]$$

Hence $\mathbf{a}^{*\mathrm{T}}\bar{\mathbf{d}} = 0.4658$, so that $\mathscr{T}_4^2(13) = 21.8016$ and, using Equation (7.26)

$$\mathscr{F} = \frac{13 - 4 + 1}{4.13} \mathscr{T}^2 = 4.19; \quad F_{.05}(4,10) = 3.48 \qquad (P = 0.03)$$

Thus we have evidence that the means are different.

Suppose we now wish to ask if the variates 'Chest circumference' and 'MUAC' provide additional information against H_0 to that supplied by 'Height' and 'Age'. We proceed by calculating the T^2-statistic using 'Height' and 'Age' only, obtaining $\mathscr{T}_2^2(13) = 6.2291$.

Now when d.f. for the hypothesis is 1,

$$1 + \frac{\mathscr{T}_p^2(f)}{f} = \frac{1}{\Lambda_{p,1,f}}$$

(see Exercise 8.3).

Hence we calculate $\Lambda_{4,1,13} = \dfrac{1}{1 + 21.8016/13} = 0.3735$ and

$$\Lambda_{2,1,13} = \frac{1}{1 + 6.2291/13} = 0.6761$$

so that the test statistic for the test of additional information is

$$\Lambda_{2.2,1,11} = \frac{0.3735}{0.6761} = 0.5524$$

Using Equation (8.19) with $p = 2$, $r = 11$, we obtain

$$\frac{1 - \Lambda}{\Lambda} \frac{11 - 2 + 1}{2} = 4.05; \qquad F_{.05}(2,10) = 4.10 \qquad (P = 0.051)$$

This is almost significant at the 5 per cent level so that we may assume that chest circumference and MUAC do provide some information against H_0 independently of the information against the hypothesis provided by height and age.

However, age is better treated as a covariate. Let us compare the means of boys and girls for the three-dimensional variate **X**, where $\mathbf{X}^T = [$Chest circumference, MUAC, Height$]$ adjusted for the one-dimensional variate, Age. The test statistic for the MANOCOVA is

$$\Lambda_{3.1,1,12} = \frac{\Lambda_{4,1,13}}{\Lambda_{1,1,13}}$$

$\Lambda_{4,1,13} = 0.3735$ is already calculated.

$\Lambda_{1,1,13} = \dfrac{1}{1 + \mathscr{T}_1^2(13)/13}$ where $\mathscr{T}_1^2(13) = \dfrac{54}{15}(-1)\left(\dfrac{68.00}{13}\right)^{-1}(-1) = 0.6882$

(the square of the univariate t-statistic for age)

so that $\Lambda_{1,1,13} = 0.9497$.

Hence $\Lambda_{3.1,1,12} = \dfrac{0.3735}{0.9497} = 0.3933$.

Using Equation (8.18),

$$-\left(12 - \frac{3 - 1 + 1}{2}\right)\ln \Lambda = 9.80; \qquad qh = 3; \qquad \chi^2_{.05}(3) = 7.82 \qquad (P = 0.02)$$

We therefore have quite strong evidence to reject the null hypothesis that the populations of boys and girls have the same mean once the samples have been adjusted for age. Note that the evidence against the null hypothesis is stronger here than in the earlier test when age was regarded as a random component.

Suppose we were now to ask if chest circumference and MUAC provide information against H_0 additional to that provided by height *after all the variates have been adjusted for age*. The test statistic is

$$\Lambda_{(2.1).(1.1),1,11} = \frac{\Lambda_{3.1,1,12}}{\Lambda_{1.1,1,12}} = \frac{\Lambda_4/\Lambda_1}{\Lambda_2/\Lambda_1} = \frac{\Lambda_4}{\Lambda_2} = \Lambda_{2.2,1,11}$$

This is just the statistic for testing the additional information against H_0 supplied by chest circumference and MUAC to that supplied by height and age which we obtained earlier (see Exercise 9.6).

One other point of especial interest concerning this example is that because of the connection between T^2 and the Fisher linear discriminant function† for two normal populations, the test of additional information is testing whether or not the exclusion of a subset of the variates results in a discriminant which is significantly less effective. (See also Rao, 1973, specifically Equation 8c.4.10 and the example in 8d.1.) □

9.6 A test of an assigned subset of linear compounds

Let $\mathbf{U} = C^T\mathbf{X}$ be a non-singular linear transformation of the response, so that if $\mathbf{U}^T = [\mathbf{U}_1^T, \mathbf{U}_2^T]$, $\mathbf{U}_1$ is a vector whose elements are q assigned linear compounds. We would then wish to test whether or not $\mathbf{U}_2$ provides information against H_0, a null hypothesis of interest, additional to that provided by $\mathbf{U}_1$. The test statistic is

$$\Lambda_{(p-q),q,h,f-q} = \Lambda_p/\Lambda_q$$

where h is the d.f. for H_0.

Because of the invariance property of Λ under non-singular linear transformations of the response, Λ_p is just that value calculated using the original variates. Λ_q may be calculated directly as

$$\Lambda_q = \frac{|C_1^T R C_1|}{|C_1^T(H+R)C_1|} \tag{9.52}$$

where H is the SSPM matrix for H_0 and $\mathbf{U}_1 = C_1^T\mathbf{X}$.

Example 9.5. In analysing the data of Table 8.3 we calculated $\Lambda_{3,2,12} = 0.2617$ in testing the null hypotheses of no temperature effects (see Section 8.3.5).

Let us see whether the variates $U_1 = X_2 - X_1$ (weight gain) and $U_2 = X_3$ are significantly less efficient for testing the null hypothesis than the given three-dimensional response (see Exercise 8.6).

Taking $H = \text{SSPM(Temperature)}$ and using Equation (9.52) with

$$C_1^T = \begin{bmatrix} -1 & 1 & 0 \\ 0 & 0 & 1 \end{bmatrix}$$

we obtain $\Lambda_2 = 1.0954/2.7047 = 0.4050$.
Hence by Equation (9.50),

$$\Lambda_{1,2,2,10} = 0.2617/0.4050 = 0.6462$$

† Fisher's linear discriminant $U = k\mathbf{a}^{*T}\mathbf{X}$ where k is some convenient constant. If we choose k so that $\hat{\text{Var}}\,\text{U} = 1$, the discriminant function is the first (and only) canonical variate discussed in Section 8.6.

Using Equation (8.18),

$$-\left(10-\frac{1-2+1}{2}\right)\ln\Lambda = -10\ln\Lambda = 4.37;$$

$qh = 2$, $\chi^2_{.05}(2) = 5.99 \qquad (P = 0.11)$

The result is not quite significant so that we could infer that an analysis using U_1 and U_2 is about as efficient in testing the temperature effects as the original variates. This provides some justification for the analysis proposed in Exercise 8.6. □

Example 9.6. An important application of this test is to the testing of an assigned discriminant function. In Example 9.4 using all four variates we found

$$\mathbf{a}^{*\mathrm{T}} = [0.2187,\ 0.1505,\ -0.0397,\ -0.1601]$$

as the solution of $13\,S\mathbf{a}^* = \mathbf{d}$ (not of $S\mathbf{a}^* = \mathbf{d}$ as in Chapter 7), so that Fisher's linear discriminant function is proportional to $\mathbf{a}^{*\mathrm{T}}\mathbf{X}$.

Suppose we wish to ask if $U = X_1 + X_2 - X_3 - X_4$ would be significantly less useful as a discriminant. The test statistic is

$$\Lambda_{3.1,\,1,\,12} = \frac{\Lambda_{4,\,1,\,13}}{\Lambda_{1,\,1,\,13}}$$

where $\Lambda_{4,\,1,\,13} = 0.3735$ as before and

$$\Lambda_{1,\,1,\,13} = \frac{1}{1+\mathscr{T}^2_1(13)/13}$$

$$\text{where} \quad \mathscr{T}^2_1(13) = \frac{9\times 6}{15}\frac{(C^{\mathrm{T}}\bar{\mathbf{d}})^2}{C^{\mathrm{T}}SC} \qquad \text{where } C^{\mathrm{T}} = [1,\ 1,\ -1,\ -1]$$

$$= \frac{54}{15}\times\frac{(-2.2)^2}{201.62/13}$$

$$= 1.1235$$

giving $\Lambda_{3.1,\,1,\,12} = 0.9205$
Hence

$$\Lambda_{3.1,\,1,\,12} = \frac{0.3735}{0.9205} = 0.4058$$

Using Equation (8.19),

$$\frac{1-\Lambda}{\Lambda}\frac{12-3+1}{3} = 4.88; \qquad F_{.05}(3,\,10) = 3.71 \qquad (P = 0.024)$$

so that we would not find the assigned discriminant acceptable.

Note that the low value, 1.1235, obtained for $\mathscr{T}^2_1(13)$ (which is just univariate $t^2(13)$) indicates immediately that U is not a good discriminant and in practice there would be no point in comparing it with the optimum discriminant $\mathbf{a}^{*\mathrm{T}}\mathbf{X}$. □

Exercises

9.1 Calculate the first canonical variates for the regression of chest circumference and MUAC on age and height for the boys' data of Example 9.4. Evaluate the sample values of U_1 and V_1 and plot. Add to your plot the two regression lines of U_1 on V_1 and of V_1 on U_1. Is the correlation between the second canonical variates significant?

9.2 Carry out a multivariate regression of chest circumference and MUAC on age and height using the pooled-within-groups residual matrix separately for boys and girls from Example 9.4. What individual regressions are being fitted? Calculate the corresponding first canonical variates and plot the sample values as in Exercise 9.1.
[*Hint*: Parallel regressions.]

9.3 Analyse the data of Example 8.1 (Table 8.1):
(*a*) regarding X_3 as a covariate;
(*b*) regarding X_2 and X_3 as covariates.
Show that (*b*) is just a univariate analysis of covariance with two covariates.

9.4 In Example 8.1, does X_2 contribute significantly to the test of hypothesis on treatment means additional to the contribution of X_1 and X_3?

9.5 Work through the theory of Section 9.4.3 in detail for an experiment involving three observations on each of two treatments where a three-variate response is observed together with a two-variable covariate.

9.6 Suppose a p-variate response $\mathbf{X}$ is partitioned into $\mathbf{X}_1$, $\mathbf{X}_2$ and $\mathbf{X}_3$ of dimensions q_1, q_2 and $p - q_1 - q_2$ respectively, with a corresponding partition of the SSPM matrices in a MANOVA, so that, in particular,

$$R = \begin{bmatrix} R_{11} & R_{12} & R_{13} \\ R_{21} & R_{22} & R_{23} \\ R_{31} & R_{32} & R_{33} \end{bmatrix} \text{on } r \text{ d.f.}$$

Suppose $\mathbf{X}_3$ is regarded as a covariate and we wish to test whether or not $\mathbf{X}_2$ provides information about an hypothesis (with SSPM matrix H on h d.f.) additional to that provided by $\mathbf{X}_1$ when the data have been adjusted for the values of $\mathbf{X}_3$. Show that the test statistic is the same as that for the test for the information provided by $\mathbf{X}_2$ additional to that supplied by $\mathbf{X}_1$ and $\mathbf{X}_3$.

9.7 Find Fisher's linear discriminant function for the data of Example 9.4 in terms of the body measurements *adjusted for age*. Is it the same as the discriminant given in Example 9.6?

PART FOUR
Multidimensional scaling and cluster analysis

The next two chapters describe two types of approach which have a number of features in common. Multidimensional scaling is essentially concerned with finding a configuration of points (or individuals or objects) from information about the 'distances' between the points. Cluster analysis is essentially concerned with seeing if there are any 'natural groupings' in a set of individuals or objects.

Both approaches make little use of an underlying stochastic model, but are useful in providing an initial summary of a set of data. The methods are exploratory in nature, the idea being to generate hypotheses rather than to test them. Some people might describe this type of approach as 'data-analysis' rather than 'statistics', but Sibson (1972) argues powerfully that methods of this type should be seen as part of a 'greater statistics'.

Another feature in common to scaling and clustering is that the methods usually operate, not on a $(n \times p)$ data matrix, but rather on an $(n \times n)$ matrix whose elements compare all pairs of individuals, either by measuring 'similarity' or by measuring 'dissimilarity'. Data like this are sometimes called *proximity data.*

CHAPTER TEN

Multidimensional scaling

10.1 Introduction

Multidimensional scaling is the term used to describe any procedure which starts with the 'distances' between a set of points (or individuals or objects), or information about these 'distances', and finds a configuration of the points, preferably in a small number of dimensions, usually 2 or 3. By 'configuration' we mean a set of co-ordinate values. For example, if we are given the road distances between all pairs of English towns, can we reconstruct a map of England? The map will of course be a two-dimensional configuration.

This type of approach is very useful because many sets of data arise, directly or indirectly, not as a $(n \times p)$ data matrix, but as an $(n \times n)$ matrix whose elements compare all pairs of individuals either by measuring 'similarity' or 'dissimilarity'. In multidimensional scaling, or scaling for short, it is often convenient to talk about the 'co-ordinates of a set of points', rather than the 'observations on a set of individuals or objects'. We also note that the word 'dissimilarity' is generally used in preference to 'distance' to describe the extent to which any two points are apart. This is because the word 'distance' has a special technical meaning, as discussed in Section 10.2, in that it is used to describe any measure of dissimilarity which satisfies the metric inequality.

Given the co-ordinates of a set of points (or individuals), it is easy to calculate the Euclidean distance between each pair of points. Scaling methods work the other way round. Given information about the dissimilarities between points, scaling methods try to find the co-ordinates of the points. In particular, the smaller the dissimilarity (or the larger the similarity) between two points, the closer we would like these points to be in the resulting spatial map.

In order to reproduce the dissimilarities between n points exactly, we may need as many as $(n - 1)$ dimensions, but our aim is to see if we can find a configuration in a much smaller number of dimensions which approximately reproduces the given dissimilarities. If a configuration in only two dimensions can be found for which the derived distances closely match the given dissimilarities, this has the great advantage that the data can easily be plotted to show up the pattern of the points. In particular, if the points fall into groups or 'clusters', then this will be evident from a visual inspection. The latter procedure is just one of the many ways of tackling cluster analysis, which is covered in Chapter 11.

The two main types of scaling procedure are called *classical scaling* and *ordinal scaling* and are considered in Sections 10.3 and 10.4 respectively. The first of these two procedures is essentially an algebraic method of reconstructing the point co-ordinates assuming that the dissimilarities are Euclidean distances, although the method is robust to the situation where the distances are distorted by errors. The method was originally proposed by Torgerson (1952, 1958) and is usually called classical or metric scaling by psychologists. It was popularized by Gower (1966) under the name of *principal co-ordinates analysis* and Gower's paper is widely quoted in recent work (e.g., Everitt, 1978, Chapter 2). But the title 'principal co-ordinates analysis' seems likely to be confused with 'principal components analysis' and does not include the word 'scaling', so we prefer the title 'classical scaling'.

Because classical scaling assumes the given 'distances' to be Euclidean, it is not suitable when the observed dissimilarities are such that the actual numerical values are of little significance and the rank order of the dissimilarities is thought to be the only relevant information. An alternative technique was therefore developed by R. N. Shepard and J. B. Kruskal in the early 1960s, which only uses the *ordinal* properties of the dissimilarities. Kruskal's original papers (1964*a*, *b*) are still worth reading. Various developments and alternatives have since been suggested, such as the Guttman–Lingoes method (see Lingoes and Roskam, 1973) but the methods commonly used today are still essentially as proposed by Kruskal.

The title originally suggested by Kruskal for this type of approach was *non-metric multidimensional scaling*, but we prefer the more modern, more descriptive title of *ordinal scaling*.

Before proceeding further, we must discuss the many different measures which are available for assessing similarity and dissimilarity.

10.2 Measures of similarity and dissimilarity

Some data sets arise directly in the form of an $(n \times n)$ similarity (or dissimilarity) matrix. To take a light-hearted example, we might be interested in comparing 5 popular car models in a similar price bracket, and ask a sample of 200 people to say which two cars they think are most alike. Then the results might appear as in Table 10.1.

The (5×5) matrix is symmetric with the diagonal terms undefined. Cars 1 and 2, and cars 4 and 5 appear to be thought of as 'similar', while car 3 seems to be separate. The frequencies are a crude measure of similarity, in that the higher the frequency, the greater the perceived similarity. A similar set of data comparing American cars is analysed in the book by Green and Carmone (1970, Chapter 2).

Many other similarity tables have been reported, particularly in the

Table 10.1 Frequencies with which each pair of cars are said to be similar

	Car 1	Car 2	Car 3	Car 4	Car 5
Car 1	—	55	15	10	13
Car 2	55	—	5	12	8
Car 3	15	5	—	9	11
Car 4	10	12	9	—	62
Car 5	13	8	11	62	—

psychological literature. For example, Kruskal and Wish (1978, p. 30) describe an experiment where 18 students were asked to rate the degree of similarity between 12 countries on a scale ranging from 'very different' to 'very similar'. The resulting map was quite unlike a geographical map in that Cuba and the USA, for example, were thought of as dissimilar even though geographically close.

Measures of similarity are often called *similarity coefficients*, and are sometimes, although not necessarily, defined to lie in the range [0, 1]. For example, in Table 10.1 we could divide each frequency by the sample size to give relative frequencies, and these will lie in the range [0, 1].

Measures of similarity and dissimilarity are closely related in an inverse way. Suppose we have a function s, defined on each pair of individuals, which measures similarity. Then it is easy to derive a corresponding measure of dissimilarity, such as $d = (\text{constant} - s)$. For example, in Table 10.1, $d = (1 - \text{relative frequency})$ is one way of measuring the dissimilarity between cars. Formally, the functions s and d are mappings from the set of all pairs of individuals into one-dimensional Euclidean space.

In many cases the measures of similarity (or dissimilarity) are not observed directly but are obtained from a given $(n \times p)$ data matrix. Given observations on p variables for each of n individuals or objects, there are many ways of constructing an $(n \times n)$ matrix showing the similarity or dissimilarity of each pair of individuals. Perhaps the most familiar measure of dissimilarity is *Euclidean distance*, such that

$$\begin{aligned} d_{rs} &= \text{distance from individual } r \text{ to individual } s \\ &= \left\{ \sum_{j=1}^{p} (x_{rj} - x_{sj})^2 \right\}^{1/2} \end{aligned} \qquad (10.1)$$

But, as we shall see later, there are many other possible measures.

Before considering these alternative measures, it will be useful to consider certain conditions which it may be desirable for measures of dissimilarity to satisfy. Jardine and Sibson (1971) define a class of dissimilarity functions, called *dissimilarity coefficients* (DC's), which are required to satisfy the following conditions: if d_{rs} denotes the dissimilarity of individual s from individual r, then:

(i) $d_{rs} \geq 0$ for every r, s
(ii) $d_{rr} = 0$ for every r
(iii) $d_{rs} = d_{sr}$ for every r, s

Conditions (i) and (ii) are straightforward. The symmetry condition in (iii) may seem obvious but is not always satisfied. For example, if we measure the dissimilarity between English towns by the time taken to travel between them, then the road system may make the journey one way longer than the other. A second example of asymmetry occurs later in Example 10.2. Tables of dissimilarities should at least be roughly symmetric, and can easily be made exactly symmetric by averaging each pair of values of the form d_{rs} and d_{sr}.

The reader may be surprised that we do *not* require a DC to satisfy the metric inequality, namely that

(iv) $d_{rt} + d_{ts} \geq d_{rs}$ for every r, s, t

Some DC's do satisfy (iv) as well as (i)–(iii) and then they are called *metrics* or *distances* rather than dissimilarities. But many DC's do not satisfy (iv) and it is by no means a necessary requirement. Note that some psychologists tend to use the word 'metric' to denote the way a variable or dissimilarity is measured without regard to whether it satisfies the metric inequality.

Euclidean distance, defined in Equation (10.1), is well known to satisfy the metric inequality as well as conditions (i)–(iii). If the data are genuine spatial co-ordinates in two or three dimensions of the same type, then Euclidean distance is the obvious measure of distance. But Euclidean distance is often less than satisfactory, particularly if the variables are measured in different units and have differing variances, and also if the variables are correlated. It is easy to construct examples where a simple change of scale not only changes the distances (obviously) but also changes the ranking of the distances. For example, suppose individuals r, s, t have weights and heights as shown in Table 10.2. Then the largest Euclidean distance is between r and s. But if we measure height in inches, then the largest distance is found to be between r and t. This seems undesirable. In order to avoid awkward changes in rankings, Euclidean distance is usually calculated for variables of different types only after each variable has been standardized to have unit variance. But even this metric does not take account of the correlation between variables.

An alternative possibility, which does take account of correlations, is to use a quantity related to the statistic $\mathcal{T}^2$, which was introduced in

Table 10.2

Individual	Weight in pounds	Height in feet
r	160	5.0
s	165	5.5
t	163	6.0

Section 7.5 for testing the difference between two vector sample means. The suggested statistic is given by

$$D_{rs}^2 = (\mathbf{x}_r - \mathbf{x}_s)^{\mathrm{T}} S^{-1} (\mathbf{x}_r - \mathbf{x}_s) \tag{10.2}$$

which may be compared with squared Euclidean distance which from Equation (10.1) is given by

$$d_{rs}^2 = (\mathbf{x}_r - \mathbf{x}_s)^{\mathrm{T}} (\mathbf{x}_r - \mathbf{x}_s) \tag{10.3}$$

There are many other possible measures of dissimilarity. For example, we can look at the angle θ between $\mathbf{x}_r$ and $\mathbf{x}_s$ and then

$$\cos^2 \theta = (\mathbf{x}_r^{\mathrm{T}} \mathbf{x}_s)^2 / (\mathbf{x}_r^{\mathrm{T}} \mathbf{x}_r)(\mathbf{x}_s^{\mathrm{T}} \mathbf{x}_s) \tag{10.4}$$

is a possible measure of similarity.

The *Minkowski metric* is a measure of distance such that

$$d_{rs} = \left\{ \sum_{j=1}^{p} |x_{rj} - x_{sj}|^R \right\}^{1/R} \tag{10.5}$$

for $R \geq 1$. When $R = 2$, this reduces to Euclidean distance. When $R = 1$, we get

$$d_{rs} = \sum_{j=1}^{p} |x_{rj} - x_{sj}|$$

which is called the city-block metric. This name arises because American cities are built in a rectangular grid arrangement and the distance travelled between two points is of the above form with $p = 2$.

Although it seems desirable at first sight to choose a measure of dissimilarity which satisfies the metric inequality, a little thought shows that this is not necessarily the case. For example, it can be shown that the statistic defined in Equation (10.2) does not satisfy the metric inequality, and yet it is often a useful measure. As a second example, consider Euclidean distance. Although this is a metric, we can easily find a monotone function of it which is not a metric. For example squared Euclidean distance is a DC, but it is easy to show by means of a counterexample that it is not a metric. Yet the rankings produced by Euclidean distance and by its square are the same and will be equally useful for many purposes. Sibson (1972) defines two DC's, d_1 and d_2, to be *global-order equivalent* if

$$d_1(w, x) \leq d_1(y, z) \Leftrightarrow d_2(w, x) \leq d_2(y, z)$$

for all individuals w, x, y, z, where the symbol $\Leftrightarrow$ means 'implies and is implied by', and $d_i(w, x)$ denotes the distance between individuals w and x using d_i. If d_1 and d_2 are global-order equivalent, d_1 and d_2 will give measures of dissimilarity which have the same rank ordering. For example, it is easy to show that Euclidean distance and squared Euclidean distance are global-order equivalent. Very often, it is ordinal properties like this which are of more interest than the metric inequality.

10.2.1 *Similarity coefficients for binary data*

One common type of data is binary data, where for example we record for each individual the presence or absence of a number of characteristics. Binary data is usually coded as 'zero' or 'one'.

Example 10.1. Construct the data matrix for the following five species – lion, giraffe, cow, domestic sheep, human being – using the following six attributes – (i) the species has a tail; (ii) the species is a wild animal; (iii) the species has a long neck; (iv) the species is a farm animal; (v) the species eats other animals; (vi) the species walks on four legs. Score one if the attribute is present, and zero otherwise. Here the species are individuals and the attributes are the variables.

The data matrix turns out to be as shown in Table 10.3. □

For binary data it is customary to begin by defining a similarity coefficient rather than a dissimilarity coefficient, and a number of different quantities have been proposed. Some of them are arranged to lie in the range [0, 1], in which case the corresponding measure of dissimilarity can be obtained by subtracting from 1.

First we can construct the (2×2) association table for each pair of individuals, as in Table 10.4. Here for example there are a attributes which are present in both individuals r and t. Note that $(a+b+c+d) = p =$ total number of attributes. Also note that in this subsection we denote the second individual by t rather than s to avoid confusion with the symbol used to denote a similarity coefficient.

Table 10.3 The data matrix

	Attribute					
	1	2	3	4	5	6
Lion	1	1	0	0	1	1
Giraffe	1	1	1	0	0	1
Cow	1	0	0	1	0	1
Sheep	1	0	0	1	0	1
Human	0	0	0	0	1	0

Table 10.4

Individual t	Individual r	
	1	0
1	a	b
0	c	d

The *simple matching coefficient* is defined by

$$s_{rt} = (a + d)/(a + b + c + d) \qquad (10.6)$$

This is the proportion of attributes for which the two individuals give the same response. Obviously $0 \leq s \leq 1$, and it can easily be shown that the corresponding measure of dissimilarity, $(1 - s)$, satisfies the conditions for it to be a DC. It can also be shown that it satisfies the metric inequality. It is customary to denote the dissimilarity $(1 - s)$ by d, even though this clashes with the customary notation for the frequency of double-zeros as used in Equation (10.6). It should always be clear from the context as to which meaning is intended. We note in passing that $(b + c)$ is called the *Hamming distance* in communication theory.

Example 10.1. (continued). For 'Lion' and 'Giraffe' we find the frequencies $a = 3, b = 1, c = 1, d = 1$, so that $s = 4/6$ and $(1 - s) = 1/3$. The complete dissimilarity matrix turns out to be as shown in Table 10.5. It is not really necessary to calculate the values of a, b, c, d for each pair of individuals. Rather we compare each line of the data matrix and then the dissimilarity d = (no. of attributes which differ)/6. □

One difficulty about the simple matching coefficient is that the presence of an attribute in two individuals may say more about the 'likeness' of the two individuals than the absence of the attribute. For example, it tells us nothing about the similarities of different subspecies in the family of wild cats to be told that neither lions nor tigers walk on two legs (except possibly in circuses!). *Jaccard's coefficient* is designed to overcome this objection and is defined by

$$s = a/(a + b + c) \qquad (10.7)$$

Thus it excludes 'double-zeros'. Then $d = (1 - s)$ turns out to be a metric DC. Jaccard's coefficient is used extensively by ecologists, for example in the comparison of plant communities in terms of the species present.

Various other coefficients have been suggested, such as

$$s = a/(a + b + c + d)$$

But in this case $(1 - s)$ is such that the dissimilarity between an individual and itself need not be zero and so $(1 - s)$ is not a DC. Despite this there are some

Table 10.5 The dissimilarity matrix

	Lion	Giraffe	Cow	Sheep	Human
Lion	—	1/3	1/2	1/2	1/2
Giraffe	1/3	—	1/2	1/2	5/6
Cow	1/2	1/2	—	0	2/3
Sheep	1/2	1/2	0	—	2/3
Human	1/2	5/6	2/3	2/3	—

circumstances where this coefficient, or some constant multiple of it, is particularly appropriate. For example Kendall (1975, p. 549) uses a similarity measure, which he denotes by K, which is equal to a. One advantage of this measure is that if we are given a data matrix X consisting of zeros and ones, then the similarity matrix containing the values of a for each pair of individuals can be obtained very easily as XX^{T}.

Some other measures are discussed by Anderberg (1973, Chapter 5).

To sum up, we remark that there is no such thing as the 'best' measure of dissimilarity. Rather, the researcher must choose the one which seems most appropriate for his particular problem. Indeed, the researcher must be prepared for situations in which there is no textbook solution, as in Example 10.2 below.

Example 10.2. In order to assess the relationships among a class of primary-age schoolchildren, their teacher asked each child to say confidentially which three other children he or she would most like to sit next to in class. The results were as shown in Table 10.6, where a 'one' indicates a selection.

For example, child number 1 selected children 5, 6 and 14 to sit next to. There should be three 'ones' in each row, but in fact child number 12 selected four

Table 10.6 The data

Child number					
1	00001	10000	00010	00000	0000
2	10000	10000	10000	00000	0000
3	10001	01000	00000	00000	0000
4	10000	11000	00000	00000	0000
5	10000	11000	00000	00000	0000
6	10000	00000	00011	00000	0000
7	00101	00100	00000	00000	0000
8	10000	01000	00010	00000	0000
9	00000	00000	00100	00000	0110
10	10100	10000	00000	00000	0000
11	01000	00000	00000	00000	0000
12	10000	11000	00000	00010	0000
13	00000	00100	00000	10100	0000
14	10000	11000	00000	00000	0000
15	10000	10000	00010	00000	0000
16	00010	00000	00100	00000	1000
17	00000	00000	00101	00001	0000
18	00000	00000	00000	11000	0001
19	01000	00000	10000	00010	0000
20	01000	00000	10000	00010	0000
21	00000	00000	00100	00000	0110
22	00000	00010	00110	00000	0000
23	00000	00010	00100	00000	1000
24	00001	00010	00000	00000	0010

children, and child 11 only selected one other child. Also note that child 19 chose herself!

This data is neither a configuration matrix, nor a distance matrix. How can we analyse it, and in particular how can we find a two-dimensional 'map' of the children to indicate the relationships between them?

The first step is to construct a matrix of 'distances' between the children. There are several ways this might be done, but we chose to find the number of steps in the 'shortest route' between each pair of children. For example, the 'distance' between child 1 and children 5, 6 and 14 was taken to be 1 in each case. Now child 1 did not select child 7 himself, but he did select child 6 and child 6 selected child 7 so the 'distance' from child 1 to child 7 was taken to be 2. Note that these 'distances', or rather dissimilarities, need not be symmetric, as the shortest route from child A to child B need not have the same length as that from B to A. If there is no possible route' between two children, the dissimilarity is set to a 'large' number, which we may take as 25, as there are 24 children and the longest possible route is of length 23. A conceptually easy (though computationally inefficient) way of finding the lengths of the shortest routes is to multiply the data matrix repeatedly by itself and see how many multiplications are needed before a positive integer, rather than a zero, appears in each position. The resulting dissimilarity matrix is shown in Table 10.7.

We shall discuss how to construct a 'map' from this matrix in Section 10.4.

Table 10.7 The dissimilarity matrix for the data of Table 10.6

Child	1	2	3	4	5	6	7	8	9	10	11	12	13	14	15	16	17	18	19	20	21	22	23	24
1	0	25	3	25	1	1	2	3	25	25	25	25	25	1	2	25	25	25	25	25	25	25	25	25
2	1	0	4	25	2	1	3	4	25	25	1	25	25	2	2	25	25	25	25	25	25	25	25	25
3	1	25	0	25	1	2	1	2	25	25	25	25	25	2	3	25	25	25	25	25	25	25	25	25
4	1	25	2	0	2	1	1	2	25	25	25	25	25	2	2	25	25	25	25	25	25	25	25	25
5	1	25	2	25	0	1	1	2	25	25	25	25	25	2	2	25	25	25	25	25	25	25	25	25
6	1	25	3	25	2	0	2	3	25	25	25	25	25	1	1	25	25	25	25	25	25	25	25	25
7	2	25	1	25	1	2	0	1	25	25	25	25	25	2	3	25	25	25	25	25	25	25	25	25
8	1	25	2	25	2	2	1	0	25	25	25	25	25	1	3	25	25	25	25	25	25	25	25	25
9	3	5	4	3	4	3	3	2	0	25	5	25	1	2	4	2	3	2	5	4	2	1	1	3
10	1	25	1	25	2	1	2	3	25	0	25	25	25	2	2	25	25	25	25	25	25	25	25	25
11	2	1	5	25	3	2	4	5	25	25	0	25	25	3	3	25	25	25	25	25	25	25	25	25
12	1	2	2	25	2	1	1	2	25	25	2	0	25	2	2	25	25	25	1	25	25	25	25	25
13	2	4	3	2	3	3	2	1	3	25	4	25	0	2	3	1	2	1	4	3	2	3	3	2
14	1	25	2	25	2	1	1	2	25	25	25	25	25	0	2	25	25	25	25	25	25	25	25	25
15	1	25	3	25	2	1	2	3	25	25	25	25	25	1	0	25	25	25	25	25	25	25	25	25
16	2	5	3	1	3	2	2	2	3	25	5	25	1	3	3	0	3	2	5	4	1	2	2	3
17	2	2	4	3	3	2	3	2	4	25	2	25	1	2	1	2	0	2	2	1	3	4	4	3
18	3	3	4	2	2	3	3	3	2	25	3	25	2	3	2	1	1	0	3	2	2	3	2	1
19	2	1	5	25	3	2	4	5	25	25	1	25	25	3	3	25	25	25	0	25	25	25	25	25
20	2	1	5	25	3	2	4	5	25	25	1	25	25	3	3	25	25	25	1	0	25	25	25	25
21	3	5	4	3	4	3	3	2	2	25	5	25	1	2	4	2	3	2	5	4	0	1	1	3
22	2	5	3	3	3	2	2	2	1	25	5	25	1	1	3	2	3	2	5	4	3	0	2	3
23	3	5	4	3	4	4	3	2	1	25	5	25	1	3	4	2	3	2	5	4	1	2	0	3
24	2	6	3	4	1	2	2	3	1	25	6	25	2	3	3	3	4	3	6	5	2	2	1	0

The type of dissimilarity used to construct Table 10.7 was exploited by Kendall (1971) and by E. M. Wilkinson and is sometimes known as the *Wilkinson metric.* □

10.3 Classical scaling

Classical scaling is an algebraic reconstruction method for finding a configuration of points from the dissimilarities between the points, which is particularly appropriate when the dissimilarities are, exactly or approximately, Euclidean distances. The approach used in this section is based on Torgerson's original work (1952, 1958).

We should note first that given a set of Euclidean distances, there is no unique representation of the points which gave rise to these distances. For example, if we know the distances between all English towns, this tells us nothing about the latitude and longitude of London. Nor does it tell us if Penzance is in the north, south, east or west, though we should be able to find that it is an 'extreme point'. Technically this means that we cannot determine the *location* and *orientation* of the configuration. The location problem is usually overcome by putting the centre of gravity of the configuration at the origin. The orientation problem means that any configuration we get may be subjected to an arbitrary orthogonal transformation (consisting of a rigid rotation plus possibly a reflection) which leaves distances and angles unchanged. In the case of the two-dimensional map of England, we might reconstruct the map 'upside down' or on its side, and we would need to rotate it and possibly look at the mirror reflection in order to get the familiar map of England. An example of the sort of map which may arise is shown later in Example 10.3(Fig. 10.1).

10.3.1 *The calculation of co-ordinate values from Euclidean distances*

If we know the exact co-ordinates of n points in p-dimensional Euclidean space, we can easily calculate the Euclidean distances between each pair of points. This can be done directly from the data matrix X, or via the $(n \times n)$ matrix $B = XX^T$, which is the matrix of between-individual sums of squares and products (the scalar products of the point vectors). The (r, s)th term of B is given by

$$b_{rs} = \sum_{j=1}^{p} x_{rj}x_{sj} \tag{10.7a}$$

The $(n \times n)$ matrix of squared Euclidean distances, D, is then such that

$$\begin{aligned} d_{rs}^2 &= \text{squared Euclidean distance between points } r \text{ and } s \\ &= \sum_{j=1}^{p} (x_{rj} - x_{sj})^2 \\ &= b_{rr} + b_{ss} - 2b_{rs} \end{aligned} \tag{10.8}$$

In this section we consider the inverse problem. Suppose we know the distances (and hence the squared distances) but not the co-ordinates. It turns out that it is easier to estimate the co-ordinates in two stages, first by finding the B matrix, and then factorizing it in the form $B = XX^{\mathrm{T}}$.

Equation (10.8) is true for all r, s. We want to invert it to get b_{rs} in terms of the $\{d_{rs}^2\}$. There is no unique solution unless we impose a location constraint. If we put the 'centre of gravity', $\bar{\mathbf{x}}$, at the origin, then $\sum_{r=1}^{n} x_{rj} = 0$ for all j. Using this result and Equation (10.7a), it is easy to show that the sum of the terms in any row of B, or in any column of B is zero. Thus summing Equation (10.8) over r, over s, and over r and s, we find

$$\begin{aligned} \sum_r d_{rs}^2 &= T + nb_{ss} \\ \sum_s d_{rs}^2 &= nb_{rr} + T \\ \sum_{r,s} d_{rs}^2 &= 2nT \end{aligned} \tag{10.9}$$

where $T = \sum_{r=1}^{n} b_{rr}$ is the trace of B. Solving Equations (10.8) and (10.9), we find

$$b_{rs} = -\tfrac{1}{2}[d_{rs}^2 - d_{r.}^2 - d_{.s}^2 + d_{..}^2] \tag{10.10}$$

where, in the usual dot notation, $d_{r.}^2$ = average term in rth row, $d_{.s}^2$ = average term in sth column, and $d_{..}^2$ = overall average squared distance. Thus, to get B from the squared distances, all we have to do is to use Equation (10.10). Computationally the easiest way to do this is to 'double-centre' D by first subtracting the average element in each row from all the terms in that row, and then doing the same thing for the columns, which automatically makes the overall average term zero. The multiplier $(-\frac{1}{2})$ can then be applied.

To get the co-ordinate matrix X from B, we simply carry out an eigenvector analysis of B. If D consists of the squares of exact Euclidean distances, then it is easy to show that B is a positive semidefinite, symmetric matrix. If B is of rank k, where $k \leq n$, then B will have k non-zero eigenvalues which we arrange in order of magnitude so that $\lambda_1 \geq \lambda_2 \geq \ldots \geq \lambda_k > 0$. The corresponding eigenvectors of unit length will be denoted by $\{\mathbf{e}_i\}$. Scale these eigenvectors so that their sum of squares is equal to λ_i. This can be done by setting $\mathbf{f}_i = \sqrt{(\lambda_i)}\mathbf{e}_i$. Then we show below that a possible co-ordinate matrix X such that $B = XX^{\mathrm{T}}$, is given by

$$X = [\mathbf{f}_1, \mathbf{f}_2, \ldots, \mathbf{f}_k] \tag{10.11}$$

Note that X is of order $(n \times k)$, so that we have a configuration in k dimensions. In other words, the co-ordinates of the rth point or individual are given by the rth components of the $\{\mathbf{f}_i\}$.

The fact that a positive semidefinite matrix can be factorized into the form XX^{T} depends on the Young–Householder factorization theorem (see

Equation (1.10)). The fact that X can take the form in Equation (10.11) depends on showing that, if X has this form, then for all $i = 1, 2, \ldots, k$,

$$(XX^{\mathrm{T}})\mathbf{e}_i = X\begin{bmatrix} \mathbf{f}_1^{\mathrm{T}} \\ \mathbf{f}_2^{\mathrm{T}} \\ \vdots \\ \mathbf{f}_k^{\mathrm{T}} \end{bmatrix}\mathbf{e}_i = X\begin{bmatrix} 0 \\ 0 \\ \vdots \\ \sqrt{\lambda_i} \\ 0 \\ \vdots \\ 0 \end{bmatrix} \text{ – (in } i\text{th row)}$$

$$= \sqrt{(\lambda_i)}\mathbf{f}_i = \lambda_i \mathbf{e}_i$$

since the $\{\mathbf{e}_i\}$ are orthonormal and $\mathbf{f}_i = \sqrt{(\lambda_i)}\mathbf{e}_i$. But $B\mathbf{e}_i = \lambda_i\mathbf{e}_i$ for all i, and so $(XX^{\mathrm{T}})\mathbf{e}_i = B\mathbf{e}_i$ for all i. Since the $\{\mathbf{e}_i\}$ form an orthonormal basis, it follows that $B = XX^{\mathrm{T}}$.

10.3.2. *The relationship between classical scaling and principal component analysis*

Classical scaling is primarily intended for analysing an $(n \times n)$ matrix of dissimilarities which are assumed to be approximate Euclidean distances. However, in this section we investigate the connection between classical scaling and principal component analysis (PCA) assuming that an $(n \times p)$ mean-corrected data matrix X is given, rather than a dissimilarity matrix. To carry out classical scaling we would calculate the $(n \times n)$ matrix of Euclidean distances and then perform the analysis described in Section 10.3.1. If X is of rank k, where $k < \min(n, p)$, this would lead to a new configuration matrix, say X^*, of order $(n \times k)$, which will generally not be the same as the original data matrix. The analysis consists essentially of finding the eigenvalues and eigenvectors of XX^{T}.

To carry out a PCA, we would find the eigenvalues and eigenvectors of the sample covariance matrix which is proportional to $X^{\mathrm{T}}X$. It is therefore natural to ask if there is any connection between the eigenvectors of XX^{T} and of $X^{\mathrm{T}}X$. If there is such a connection, then we have found an interesting link between the two techniques.

Let $\{\lambda_i\}$ and $\{\mathbf{a}_i\}$ denote the eigenvalues and eigenvectors of $X^{\mathrm{T}}X$. Then

$$(X^{\mathrm{T}}X)\mathbf{a}_i = \lambda_i\mathbf{a}_i$$

Premultiplying by X, we find

$$(XX^{\mathrm{T}})X\mathbf{a}_i = \lambda_i X\mathbf{a}_i$$

From the form of this equation, we see that the eigenvalues of XX^{T} must be the same as those of $X^{\mathrm{T}}X$, while the eigenvectors of XX^{T}, denoted by $\{\mathbf{e}_i\}$, will be related to those of $X^{\mathrm{T}}X$ by a simple linear transformation, since $\mathbf{e}_i$ must be proportional to $X\mathbf{a}_i$. Note that $\mathbf{e}_i$ is of order $(n \times 1)$, while $\mathbf{a}_i$ is of order $(p \times 1)$. Now eigenvectors are usually normalized so that their sum of squares is one, and we assume this applies to $\mathbf{e}_i$ and $\mathbf{a}_i$. But $X\mathbf{a}_i$ is a vector giving the component scores for each individual for the ith principal component (see Chapter 4), and the sum of squares of $X\mathbf{a}_i$ is known to be λ_i. Thus we must have

$$\sqrt{(\lambda_i)}\mathbf{e}_i = X\mathbf{a}_i$$

(possibly after changing the arbitrary sign of $\mathbf{e}_i$ or $\mathbf{a}_i$)
Thus, instead of calculating the component scores by transforming the eigenvectors of $X^{\mathrm{T}}X$, we could get them directly from the eigenvectors of XX^{T} by

$$\mathbf{f}_i = \sqrt{(\lambda_i)}\mathbf{e}_i \tag{10.12}$$

Thus, the results of PCA are exactly equivalent to those of classical scaling if the distances calculated from the data matrix are Euclidean. Gower (1966) has shown that this equivalence also applies to a data matrix consisting solely of zeros and ones if the 'distances' between individuals are calculated using the simple matching coefficient.

We have already remarked that it is impossible to determine a unique orientation for the configuration from the distances. The above result shows that if we carry out an eigenvector analysis of $B = XX^{\mathrm{T}}$, we get a configuration whose co-ordinate axes are the principal axes resulting from a principal component analysis. In other words, the first axis accounts for the largest possible proportion of the total variance, and so on.

The equivalence of PCA and classical scaling means that if we want to reduce the effective dimensionality of a given data matrix by transforming to new co-ordinate axes, then there is no point in carrying out both analyses. This means that, if $n > p$, then a PCA will generally be preferred because it is generally easier to find the eigenvectors of the $(p \times p)$ matrix $X^{\mathrm{T}}X$ than those of the larger $(n \times n)$ matrix XX^{T}. Surprisingly, these results are not made clear in the literature, which often makes it sound as though analysing the XX^{T} matrix will give different results from those resulting from the $X^{\mathrm{T}}X$ matrix. In particular, the plot of component scores in Fig. 4.1 can equally well be regarded as resulting from a principal component analysis or from classical scaling (or principal co-ordinates analysis).

As a final comment, we remind the reader that PCA is usually applied to the sample correlation matrix rather than the sample covariance matrix. The corresponding analysis in classical scaling is to work with standardized Euclidean distances where each variable is arranged to have unit variance.

10.3.3. *Classical scaling for a dissimilarity matrix*

Classical scaling comes into its own when we do *not* have the data matrix, but instead have an $(n \times n)$ matrix of dissimilarities (or similarities). These dissimilarities must satisfy the requirements of a dissimilarity coefficient so that the 'distance' matrix has zeros on the diagonal and symmetric, non-negative entries off the diagonal. Then we proceed exactly as if the 'distances' were known to be Euclidean. First the 'distances' are squared, then the B-matrix is formed using Equation (10.10), and finally the eigenvectors of B are found. These three steps are easy to carry out on a computer.

How do we interpret the results? If the 'distances' are exact Euclidean distances, then we get co-ordinates relative to the principal axes. But if the 'distances' are not Euclidean, then B need not be positive semidefinite. Then we will not, as in the Euclidean case, find k positive eigenvalues and $(n-k)$ zero eigenvalues. Rather there will be n eigenvalues, the last few of which may well be negative. There should always be at least one zero eigenvalue because the sum of the elements in each row of B is chosen to be zero.

If we are looking for a configuration in a given number of dimensions, say p^*, then we simply look at the p^* largest eigenvalues and their corresponding eigenvectors. Then, using Equation (10.12), we get the co-ordinate matrix as in Equation (10.11).

If we are unsure as to the dimension, then we look to see how many eigenvalues are 'large'. All negative eigenvalues are discarded, together with any positive eigenvalues which are 'small'. How small is 'small'? Similar arguments apply as in PCA. If, say, three eigenvalues are much larger than the rest, and account for a large proportion of the total variance, then we can approximately reproduce the distances by selecting a configuration in three dimensions. As a general rule, the largest negative eigenvalue should be less than the smallest positive eigenvalue which is considered to be 'large'. Another useful rule, called the *trace criterion*, says that the sum of the 'small' eigenvalues, both positive and negative, should be approximately zero, so that the sum of the 'large' positive eigenvalues will approximately equal the trace of X^TX.

It turns out that classical scaling is surprisingly robust to departures from Euclidean distance. Comparatively large random errors can be added to the Euclidean distances without seriously distorting the resulting configuration of points (see Sibson, 1979). Even if we take a monotonic function of the distances, we will probably still get a reasonable configuration.

However, if the dissimilarities are not even approximately Euclidean distances, then classical scaling is generally not advised. This situation may be indicated by finding that B has a 'large' negative eigenvalue or by finding that B has several medium-sized positive eigenvalues so that the solution tries to spread into a larger number of dimensions. In these situations we prefer a technique which uses only the rank order of the dissimilarities, and this will be described in Section 10.4. However, it turns out that classical scaling can still be useful in providing an initial configuration for the iterative method that is used there.

10.3.4. *Interpretation of results*

If a configuration in two dimensions is found, the resulting points can easily be plotted. The pattern of points which results should help us to interpret the data.

Example 10.3. Classical scaling was applied to the table of road distances

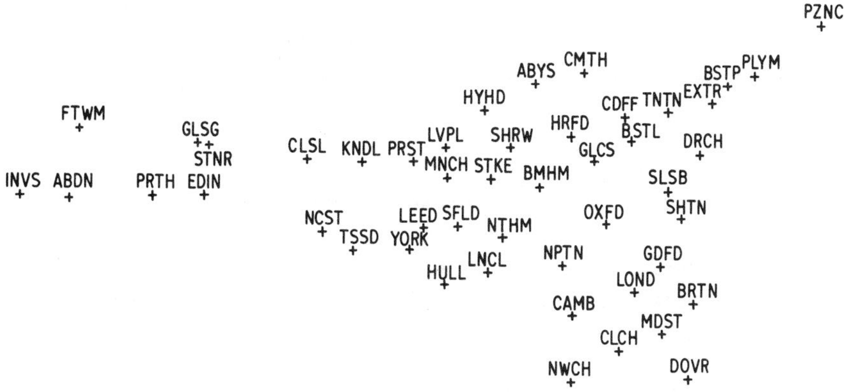

Figure 10.1 A map of Great Britain reconstructed by classical scaling.

between pairs of towns in Great Britain as given in the AA handbook. A map in two dimensions was found. The result is shown in Fig. 10.1. Most of the abbreviations should be obvious (e.g. PZNC denotes Penzance). In a sense this is a pointless exercise as we know what the map should really look like. But it does illustrate a number of useful points about classical scaling. Firstly, we note that the orientation of the map is incorrect as it is 'on its side'. Furthermore, the map has been reflected from its usual mode so that the east and west sides have been transposed.

Secondly, we note that road distances are not exact Euclidean distances though they will usually be 'close' to them. But in one or two cases (e.g., Penzance to Carmarthen), the road distance is considerably larger than the Euclidean distance because of geographical features, and this has the effect of distorting the positions of some of the extreme points. □

When interpreting a graph resulting from scaling, the above example illustrates the point that there is no reason why the axes which result should have any particular 'meaning'. Indeed, the orientation can be subjected to an arbitrary orthogonal transformation. Thus it may be possible to find new, more sensible, axes which do lend themselves to an obvious interpretation. The search for meaningful dimensions is one of the main aims of scaling, and Kruskal and Wish (1978, Chapter 2) suggest that multiple linear regression between some direction in the configuration and various characteristics of the points may be useful in trying to interpret the dimensions.

Another obvious way the graph can be used is to look for groups or clusters of points which are close together. Indeed, it is sometimes useful to draw straight lines linking all those points whose dissimilarity is less than some carefully chosen threshold value.

If a configuration in three dimensions is found, the problems of interpretation are considerably more difficult. It may be helpful to plot the dimensions two at a time and multiple linear regression may also be helpful. An example is given by Kruskal and Wish (1978, p. 37).

The number of successful published examples of classical scaling (or principal co-ordinates analysis) is rather small, partly because the method is not widely understood and partly because many observed dissimilarities do not approximate to Euclidean distances, so that ordinal scaling has to be used.

10.3.5. *Some related methods*

We have already remarked that Gower's method of principal co-ordinates analysis is equivalent to classical scaling, although this is not immediately obvious on reading Gower's original 1966 paper. Gower operates on a similarity matrix rather than a dissimilarity matrix, finds its eigenvectors, carries out a principal component analysis of the resulting co-ordinates, and then shows that these two stages can be collapsed into one in order to give what we call the matrix B. We believe our approach to be rather more straightforward.

It is also worth introducing the term *metric multidimensional scaling*, which, as used for example by Kruskal and Wish (1978, p. 22) is rather more general than classical scaling. In the latter, we assume that the observed dissimilarities are approximately Euclidean. In order to check this assumption it is useful to plot the observed dissimilarities against the derived Euclidean distances which result from the derived configuration, and then we should get a relationship which is approximately linear and which goes through the origin. But sometimes this does not happen. If the relationship is not linear or does not go through the origin, then some alternative model should be tried. For example, dissimilarity data sometimes arise in psychology which are thought to be Euclidean apart from an additive constant. This situation was originally discussed by Torgerson (1958) and gives rise to what is called 'the additive-constant problem'. More generally, we may want to make some sort of transformation of the observed dissimilarities before carrying out classical scaling. The term *metric multidimensional scaling* appears to be used when there is some specified analytic function relating the observed dissimilarities and the derived distances. However, if this function is merely known to be monotonic, then ordinal scaling is appropriate as discussed in the next section.

10.4. Ordinal scaling

In ordinal scaling, as in classical scaling, our basic aim is to find n points whose interpoint distances are in good agreement with the given dissimilarities between n objects or individuals. In particular, we would like to represent the data, without too much distortion, in only two or three dimensions if possible. The basic difference from classical scaling is that we use only the rank order of the dissimilarities.

The reason for pursuing methods like this is that in many situations, particularly in psychology, the measures of similarity or dissimilarity which arise are such that their exact numerical values have little intrinsic meaning. Instead, we are interested in the ranking of the values. For, example, if we consider the dissimilarities in Table 10.7, we see that the dissimilarity of child 2 from child 1 is 25, while the dissimilarity of child 3 from child 1 is only 3. We can certainly say that the first dissimilarity is larger than the second, but there is not much more we can say as the dissimilarities, 25 and 3, are not familiar Euclidean distances. In particular, it would be inadvisable to regard the first dissimilarity as being $8\frac{1}{3}$ times as large as the second dissimilarity.

As in classical scaling, we cannot uniquely determine the location and orientation of the resulting configuration, and this should be borne in mind by the researcher. A further problem in ordinal scaling is that, as we only use the ordinal properties of the dissimilarities, there is no implied scale for the configuration. This is overcome by putting the centroid of the configuration at the origin and then requiring that the root-mean-square distance from the points to the origin is unity.

Let us denote the given dissimilarities by $\{\delta_{rs}\}$, as this is the standard notation in ordinal scaling. Suppose we want to find a configuration in p^* dimensions. We use p^*, rather than p, to emphasize that the number of dimensions is not given. In practice, several 'low' values of p^*, such as 2, 3 and 4, may be tried. Two dimensions is particularly popular, as the configuration may then be plotted on a graph. For technical reasons, it is not generally recommended that ordinal scaling should be used to find a configuration in just one dimension. Yet in some situations a one-dimensional solution is required. For example, in archaeological seriation problems, we may have observations resulting from excavations at different sites and want to put the sites into temporal order. The recommended approach here is to construct suitable measures of dissimilarity, find a two-dimensional solution and then see if the points fall into a comparatively long narrow band, which need not necessarily be straight. If they do, then a one-dimensional solution can be inferred in an obvious way. (Note that a completely different approach to seriation, which is appropriate for analysing contingency tables including presence/absence data, is provided by correspondence analysis – see Hill, 1974 and Greenacre, 1984).

10.4.1 *The iterative procedure*

The general procedure adopted in ordinal scaling is as follows. No analytic solution is possible. Instead, we have to use an iterative approach. We begin by 'guessing' an initial configuration in p^* dimensions, calculate the Euclidean distances between each pair of points in the guessed configuration and compare these derived distances with the observed dissimilarities. The derived distances are customarily denoted by $\{d_{rs}\}$ and should not be confused with the given dissimilarities, $\{\delta_{rs}\}$. If the $\{d_{rs}\}$ are in the same rank order as the $\{\delta_{rs}\}$, then we have found a satisfactory configuration. But in practice it will usually be impossible to find any configuration which gives a perfect match between the d's and the δ's. Instead, we must find a configuration such that the ordering of the d's is 'as close as possible' to that of the δ's.

The goodness-of-fit of any guessed configuration is found by constructing a

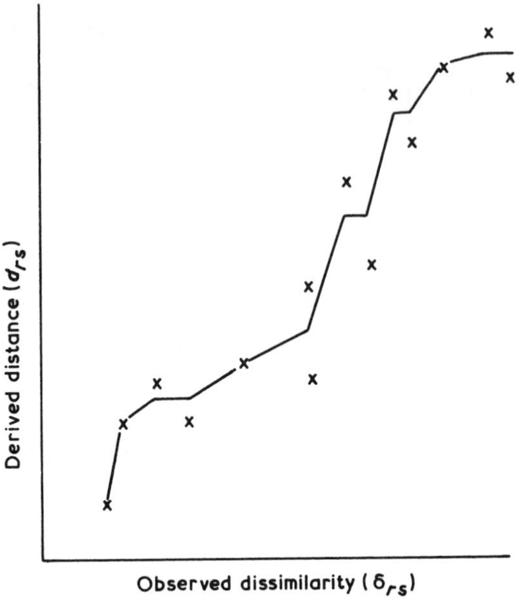

Figure 10.2. An example of a least-squares monotone regression line.

monotone least-squares regression relationship between the d's and the δ's. Let $\hat{d}_{rs}$ denote the point on the curve corresponding to δ_{rs}. The usual monotone requirement says that, if $\delta_{rs} < \delta_{tu}$, then we must have $\hat{d}_{rs} \leq \hat{d}_{tu}$. Note particularly that equality is allowed in the relationship between $\hat{d}_{rs}$ and $\hat{d}_{tu}$, but not in that between δ_{rs} and δ_{tu}. This monotone requirement, sometimes called the *primary* monotone condition, does not say what to do in the event of *ties* between the δ's. In fact, if $\delta_{rs} = \delta_{tu}$, then we can have $\hat{d}_{rs}$ either less than, equal to, or greater than $\hat{d}_{tu}$. A more restrictive monotone requirement, called the *secondary* approach, additionally requires that if $\delta_{rs} = \delta_{tu}$, then $\hat{d}_{rs} = \hat{d}_{tu}$. But this is usually too restrictive and often leads to convergence problems. Thus it is not generally recommended. Note also that the (primary) monotone requirement considers the ranks of *all* the dissimilarities, and this is sometimes referred to as *global scaling*. In some practical situations, it seems more reasonable to try an alternative procedure called *local scaling*, where comparisons are only made between dissimilarities with a common end-point, but we shall say no more about this here.

Given values of $\{\delta\}$ and $\{d\}$, an efficient terminating algorithm exists for finding the least-squares monotone regression curve, but we will not describe it here. An example of such a curve is shown in Fig. 10.2. In this example there is reasonably good, though not perfect, agreement between the order of the d's and δ's, indicating that the corresponding configuration is 'quite good', though probably not optimal. The monotone regression curve is different in kind from familiar regression curves, as its derivative is not continuous. In fact the curve is really only defined at the given values of δ, but it is helpful to draw in the complete line for pictorial clarity. If at some point the d's decrease as the δ's increase, then the monotone curve is horizontal. The least-squares requirement ensures that the sum of squared (vertical) deviations from the curve is as small as possible subject to the monotone requirement. For example, let us consider the four smallest δ's in Fig. 10.2. The $\hat{d}$-values for the first two points are equal to the d-values, but for the third and fourth points where the d-values decrease, the $\hat{d}$-values are constant and equal to the average of the two d-values.

In order to assess how well the derived configuration matches up to the given dissimilarities, we can calculate a statistic, called the *stress*, which is defined by

$$S = \sum_{r<s} (d_{rs} - \hat{d}_{rs})^2 / \sum_{r<s} d_{rs}^2 \qquad (10.13)$$

This is simply the residual sum of squares, normalized so that it will usually lie in the range [0, 1]. The denominator in Equation (10.13) is the normalizing factor, and must be included as no unique scale can be inferred from the δ's. The stress is often multiplied by 100 and expressed as a percentage. If there is a perfect monotone relation between the d's and the δ's, the stress will be zero. Kruskal suggested that a stress of 0.05 (or 5 per cent) is 'good', while a stress of 0.20 (or 20 per cent) is 'poor', but these guidelines are now known to be too simple as the stress depends on n and p^*. The stress will tend to increase if n increases and/or p^* decreases. Some researchers prefer to normalize the stress in Equation (10.13) by dividing by $\Sigma(d_{rs} - \bar{d})^2$ rather than Σd_{rs}^2, as it is claimed that this is less likely to give degenerate solutions for some types of data. But it is also more likely to give local optimum problems.

After trying an initial configuration, we move the points around so as to try to reduce the stress. As S is explicitly differentiable, an iterative procedure can be used, similar to that used in 'hill-climbing', or rather 'the method of steepest descent', as we are trying to find a minimum. Unfortunately, sophisticated hill-climbing procedures do not work very well, as the objective function is often nowhere near quadratic, so that the methods used today are still similar to those originally proposed by Kruskal. We will not describe them here as computer programs are readily available and the problems are related more to numerical analysis.

One problem that does need to be mentioned is that there may be local minima where no small change in the configuration will reduce the stress but where the configuration is not 'best' overall. Then there is no certainty that the iterative procedure will reach a global minimum. This problem is usually overcome by trying several different initial configurations. The configuration in the first p^* dimensions which results from classical scaling is one good choice for an initial configuration. It is also worth knowing that the stress function will not always converge, in which case the maximum allowed number of iterations will be exceeded.

The dimension p^* is usually unknown. One way of assessing the dimensionality is to find the minimum stress in say two, three and four dimensions and to plot this minimized stress against p^*. The stress should always decrease as the dimensionality increases. But the results given by Kruskal and Wish (1978, Chapter 3) suggest that it may be possible to assess the dimensionality by looking for an 'elbow' in the graph of stress versus p^* beyond which the stress decreases relatively slowly, provided that the error is relatively small and provided that n is larger than about 20.

The complicated nature of the iterative routine means that the amount of computing involved in ordinal scaling is much greater than that involved in classical scaling. For a typical set of data, ordinal scaling might take between 5 and 15 times as long as classical scaling.

10.4.2. *Interpreting the results*

Having assessed the 'best' value of p^* and the resulting optimal configuration, we then examine the configuration in the same way as described in Section 10.3.4. In particular, if $p^* = 2$, we can easily plot a 'map' of the data.

There are several successful examples of ordinal scaling in the literature. Apart from the psychological literature, the reader is referred particularly to the published work of D. G. Kendall and R. Sibson. Examples include the reconstruction of maps from archaeological data, the seriation of a set of graves at Münsingen from artifact data, and the reconstruction of a map of France using data which simply showed whether or not each pair of 'départements' were touching. In the latter case, the data were converted to dissimilarities by using the Wilkinson metric which effectively finds the smallest number of départements between each pair (see the first part of Example 10.2; and Kendall, 1975). Further examples from the behavioural sciences are given by Romney *et al.* (1972), and from archaeology by Hodson *et al.* (1971).

Example 10.2. (continued). The data in Table 10.7 show the dissimilarities between all pairs of children in the given class. It would be useful to have a pictorial representation of the relationships between the children, and a two-dimensional scaling solution seems to be called for. The dissimilarities are such that only the rank order seems relevant. Thus there seems little point in trying classical scaling. Ordinal scaling was carried out and the resulting configuration in two dimensions is shown in Fig. 10.3.

There is one point of note in the analysis. The given dissimilarity matrix is not symmetric. One way to proceed is to construct a new symmetric dissimilarity matrix with elements δ^*_{rs} such that

$$\delta^*_{rs} = \delta^*_{sr} = \tfrac{1}{2}(\delta_{rs} + \delta_{sr})$$

and simply carry out ordinal scaling on this new matrix. Alternatively, we can carry out ordinal scaling on the non-symmetric matrix, where the summations in Equation (10.13) will now be over all possible pairs of (r, s) and not just those where $r < s$. This requires more computation. The latter approach was used in this case.

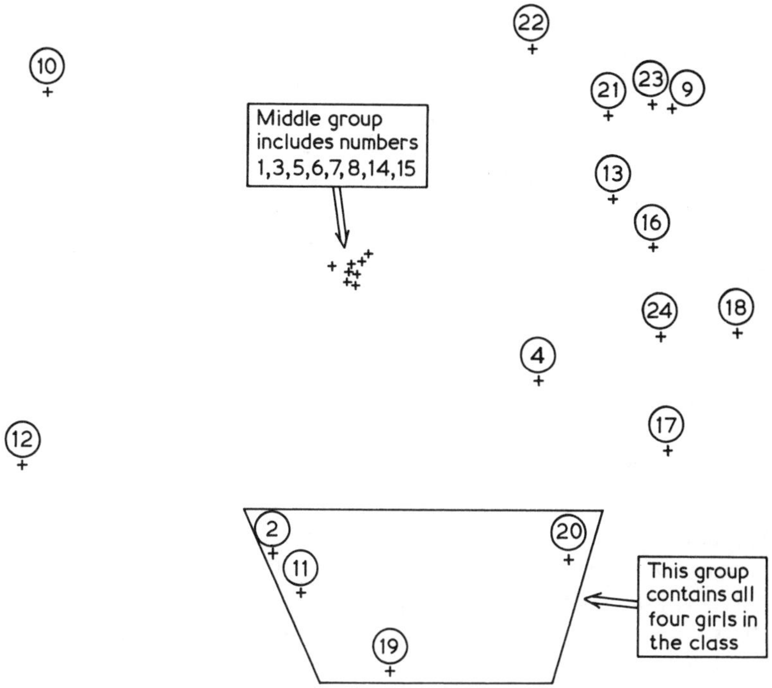

Figure 10.3. A two-dimensional representation of 24 children.

The interpretation of Fig. 10.3 is surprisingly clear-cut. The four girls in the class come out in a group towards the bottom of the graph. The tight cluster of eight children in the middle of the graph consists of eight boys who their teacher judged to be the most intelligent. The two isolated points on the left of the graph are the two children who were not selected by anyone else in the class. Of course, it could be argued that all this information could be obtained by a close study of the original data matrix in Table 10.6, but ordinal scaling certainly makes it a lot easier. □

10.5 A comparison

A number of experiments have been carried out at Bath University to compare ordinal scaling (OS) and classical scaling (CS). This was done by taking a known configuration of points, subjecting the resulting Euclidean distances between pairs of points to a variety of transformations and error disturbances, and then attempting to reconstruct the original configuration using OS and CS. The resulting configurations were translated, rotated, scaled, and possibly reflected in order to make them as close as possible to the original configuration. Then the success of the reconstruction procedure was measured by

$$\sum_{\text{all points}} \begin{pmatrix} \text{squared Euclidean distance from actual position} \\ \text{to achieved position} \end{pmatrix}$$

after normalizing by dividing by the trace of $X^{\mathrm{T}}X$, where X is the original configuration matrix. It was found that OS gave more or less the same results as CS when distances are approximately Euclidean, but that OS gives satisfactory results when CS gives poor results in cases where distances are not Euclidean. In other words, OS is generally better than CS. It was further found that even better results could generally be obtained by a third scaling procedure, called *least-squares* scaling, where we choose the configuration so as to minimize $\sum$(achieved distance − observed dissimilarity)^2, where the summation is over all pairs of points. This procedure is conceptually simple, but iterative procedures must be used to find the 'best' configuration. Good results were also obtained from a technique called *non-linear mapping* (see Everitt, 1978, Chapter 2) which is closely related to least-squares scaling. Here the configuration is chosen so as to minimize $\sum$[(achieved distance − observed dissimilarity)^2/observed dissimilarity].

10.6 Concluding remarks

In an introductory chapter like this, it is impossible to describe fully the many recent developments in the subject. One important advance is *individual differences scaling* or INDSCAL (see Carroll and Chang, 1970) which is applied to 'three-way data' where there are at least two dissimilarity matrices, and the analysis takes account of systematic differences between them. The reader is referred to Kruskal and Wish (1978) and Springall (1978) for a review of recent developments.

It is noticeable that the use of scaling has shifted away from being exploratory and 'data-analytic' to being 'confirmatory'. The use of explicit models is now more widespread. For example, significance tests have been proposed, based on the stress function, for assessing the 'true' number of dimensions, assuming that this exists. However, some of these recent developments require the user to make a number of possibly unrealistic assumptions about the data, and we view recent trends with some reserve. We regard multidimensional scaling as an essentially exploratory tool.

Exercises

10.1 Observations on two variables are made for five individuals as follows:

Individual	Variable	
	1	2
1	1	1
2	1	2
3	6	3
4	8	2
5	8	0

Construct a dissimilarity matrix using (*a*) squared Euclidean distance; (*b*) the city block metric.

10.2 Show that the city-block metric satisfies the metric inequality. The values of three variables are measured for each of six individuals as follows:

Individual	Variable 1	2	3
1	2	4	9
2	5	6	6
3	8	8	7
4	6	5	6
5	2	5	9
6	8	9	5

Construct the distance matrix for the six individuals using the city-block metric.

10.3 If s_1, s_2 denote the simple matching coefficient and Jaccard's coefficient respectively, show that $d_1 = (1 - s_1)$ and $d_2 = (1 - s_2)$ are both metric dissimilarity coefficients (DC's).
But show that $d_3 = (1 - s_1)^2$ and $d_4 = 1/(s_1 + 1)$ are not metric DC's.

10.4 Show that the two dissimilarity coefficients (DC's) given by $d_1 =$ Euclidean distance and $d_2 =$ squared Euclidean distance are global-order equivalent (GOE – see Section 10.2).

Show by means of a counter example that the following pairs of DC's are not GOE:
(*a*) Euclidean distance and the city-block metric.
(*b*) Euclidean distance in one set of co-ordinates and Euclidean distance after one of the variables has been scaled.

10.5 Two DC's, d_1 and d_2, are said to be *local-order equivalent* (LOE) if $d_1(w, x) \leq d_1(w, z) \Leftrightarrow d_2(w, x) \leq d_2(w, z)$ for every set of individuals w, x, z. Clearly, global-order equivalence implies local-order equivalence, but the reverse is not necessarily true. Try to construct two distance matrices for four individuals which are LOE but not GOE (if you get stuck, an example is given by Sibson, 1972, p. 318).

10.6 The values of five binary variables are measured for each of five individuals as follows:

Individual	Variable 1	2	3	4	5
1	1	1	1	0	1
2	0	0	1	1	1
3	1	1	1	1	1
4	0	1	0	1	0
5	0	0	1	1	1

Construct a dissimilarity matrix for the five individuals using:
(*a*) the simple matching coefficient; (*b*) Jaccard's coefficient.

CHAPTER ELEVEN
Cluster analysis

11.1 Introduction

The basic aim of cluster analysis is to find the 'natural groupings', if any, of a set of individuals (or objects, or points, or units, or whatever). This set of individuals may form a complete population or be a sample from some larger population. More formally, cluster analysis aims to allocate a set of individuals to a set of mutually exclusive, exhaustive, groups such that individuals within a group are similar to one another while individuals in different groups are dissimilar. This set of groups is usually called a *partition*.

When only two variables are measured on each individual, it is relatively easy to plot the data and pick out clusters by eye. The 'obvious' partition for the bivariate data plotted in Fig. 11.1 consists of the three groups which are encircled.

The groups forming a partition may be subdivided into smaller sets or grouped into larger sets, so that one eventually ends up with the complete hierarchical structure of the given set of individuals. This structure is often called a *hierarchical tree* and can be represented diagrammatically as in Fig. 11.2.

Many clustering methods are more concerned with finding the hierarchical structure of a set of individuals than with finding a single partition, and it is helpful to distinguish between these two objectives, though there is inevitably some overlap. In particular, we can always get a partition from a tree by drawing a horizontal line through the tree at an appropriate point. This is sometimes called 'chopping the tree'. Suppose we want a partition of g clusters in a set of n individuals. Then we move the horizontal line down the tree until it cuts g vertical lines. Then we simply group the individuals that are connected to the same vertical line. However, this may not necessarily be the best way of finding such a partition. By varying g, we note that the hierarchical structure implicitly implies a partition for every possible number of groups from 1 to n.

Classification is one of the basic aims of all scientific research and is something that the human brain does automatically in most aspects of life. For example, the material in this book is partitioned into chapters, and hierarchically structured into parts, chapters, sections and subsections.

Many authors use the term 'cluster analysis' synonymously with 'classifi-

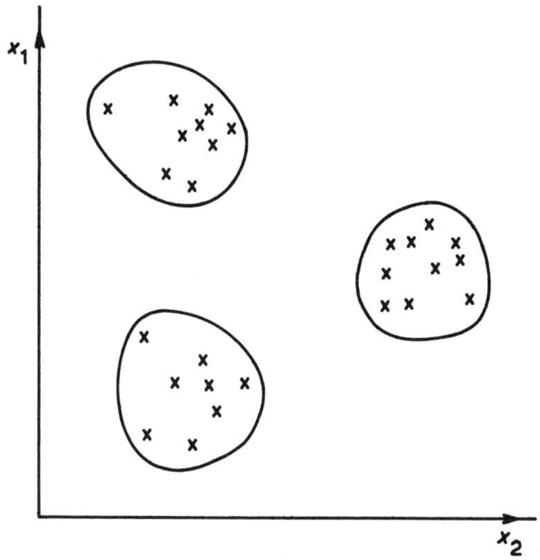

Figure 11.1. A scatter diagram of bivariate data where three clusters are evident.

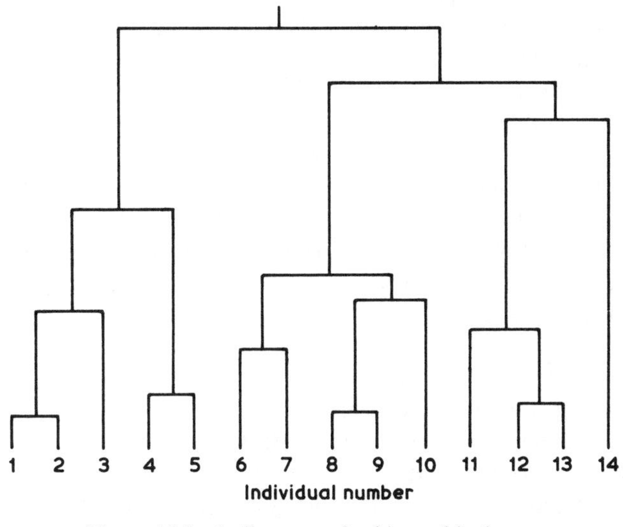

Figure 11.2. A diagram of a hierarchical tree.

cation' and with 'taxonomy', the latter word being the biologist's term for classification. However, some authors (e.g., Anderberg, 1973; Gnanadesikan, 1977) use the term 'classification' to mean the assignment of individuals to a *pre-established* set of categories. This type of problem, which is related to discriminant analysis, is more usually called *assignment*, *identification* or *diagnosis*. We shall use 'classification' synonymously with 'cluster analysis' to denote grouping techniques where the categories are determined from the data, and will not consider the assignment problem in this chapter at all.

In order to carry out a cluster analysis we need to measure the *similarity* (or dissimilarity) of every pair of individuals. As described in Section 10.2, there are many ways of doing this. The similarities are sometimes observed directly, while in other cases they are derived from the data matrix in an appropriate way. Standardized Euclidean distance is one of the most common measures of dissimilarity.

There is a large literature on cluster analysis, including several specialized books. Anderberg (1973) and Everitt (1974) are both suitable for the beginner, while Hartigan (1975) provides detailed algorithms and computer programs. Jardine and Sibson (1971) provide a proper mathematical treatment of the subject. The classic text by Sokal and Sneath (1963) was the first important book on the subject and is of some historical interest. Various examples are given in the above books as well as in Hodson *et al.* (1971) and in many other books and articles.

Cluster analysis was particularly popular in the 1960s, but in recent years has found rather less favour, partly because of an important review article by Cormack (1971), who adopted a somewhat sceptical attitude to the whole area. The first sentence of Cormack's paper is worth repeating, namely: 'The availability of computer packages of classification techniques has led to the waste of more valuable scientific time than any other "statistical" innovation (with the possible exception of multiple-regression techniques).' Cormack also deplored the tendency to regard cluster analysis as a satisfactory alternative to clear thinking. Most (though not all!) users are now well aware of the drawbacks to cluster analysis and of the precautions which need to be taken to avoid being faced with irrelevant or misleading results. A recent review is given by Everitt (1979). Despite some reduction in research activity in this area, the literature is so large, and the number of available clustering methods is so high, that we shall only be able to describe some of the more important ones in this introductory chapter.

11.1.1 *Objectives*

A variety of objectives have been suggested for cluster analysis. They include (*a*) data exploration; (*b*) data reduction; (*c*) hypothesis generation; and (*d*) prediction based on groups.

The first objective is self-explanatory. One often starts a cluster analysis with little clear idea as to how a set of individuals is structured, so that the method may be seen as a 'tool of discovery'. If the clusters turn out to be compact, then we may be able to reduce the information on n individuals to information about a much smaller number of groups. Hopefully the analysis may generate some ideas (or hypotheses) regarding the structure of the population. In particular, the number of clusters suggested by the first analysis can be treated as an hypothesis to be tested by a new set of data. The groupings produced by a cluster analysis may sometimes be used in later predictive studies.

Broadly speaking, we can say that a cluster analysis has been successful if it brings to light previously unnoticed groupings in a set of data or helps to formalize its hierarchical structure.

11.1.2 *Clumping, dissection and clustering variables*

This section briefly considers three topics which are closely related to cluster analysis but about which we shall say little explicitly. The topics are clumping, dissection and the clustering of variables.

The term *clumping* is usually applied to clustering methods where the groups are allowed to overlap. For example, in trying to classify words according to their meaning, we sometimes find that a word has two or more distinct meanings and needs to be allocated to more than one group. The reader is referred to the section on overlapping clusters in Cormack (1971).

The term *dissection* is used when one has a single homogeneous population, so that there is no natural grouping, and yet one still wants to split the population into subgroups. For example, although London is a more or less continuous collection of houses, the Post Office still finds it administratively convenient to divide the city into postal districts. In this sort of situation, the number of subgroups is clearly arbitrary as is the method of obtaining them. Marriott (1974, p. 60) points out that one simply chooses the most convenient practical method which is appropriate for the given situation. No statistical theory is involved.

The third topic we shall say little about explicitly is that of grouping *variables* rather than individuals. The objective in clustering variables is to see if we can find subsets of variables which are so highly correlated among themselves that we can use any one of them, or some average of them, so as to represent the subset without serious loss of information (see Kendall, 1975, p. 32).

In view of the duality between measurements on variables and on individuals, it turns out that many of the techniques described in this chapter can also be applied to clustering variables. In order to do this we need to have a similarity (or dissimilarity) between every pair of variables. The 'obvious' measure of similarity is some sort of correlation coefficient, since if two variables are highly correlated they may be regarded as giving similar effects. Other measures of similarity for variables are discussed by Anderberg (1973, Chapter 4).

We have in effect already considered some alternative ways of clustering variables, as this is one of the aims of principal component analysis and of factor analysis. However, as noted in the summary of Chapter 4 and by Kendall (1975, Section 3.6), it is often just as easy to group variables by visually inspecting the correlation matrix. If we combine variables having a high correlation (or similarity), then we are effectively carrying out a crude form of cluster analysis.

11.1.3 *Some other preliminary points*

An immediate difficulty in cluster analysis is that there is no completely satisfactory way of defining a 'cluster', although we usually have an intuitive idea as to what is meant by the word. For example, we want the clusters to be parts of the p-space where points are 'densely' packed, but which are separated by parts of low density. Put another way, we want the clusters to be internally 'coherent' but to be isolated from other clusters.

Users of clustering methods may initially be surprised to find that different methods applied to the same set of data will often produce structures which are substantially different. This is because the choice of a clustering method implicitly imposes a structure on the population and is often tantamount to defining a cluster.

Another problem is that it simply may not be possible to classify the individuals in any useful way. Nevertheless, a hierarchical clustering method will still find a tree and a partitional clustering method will still find a partition even though the clusters may be quite spurious. The ability of clustering methods to detect non-existent clusters is well established.

If a classification does exist, a further problem is that the data may admit more than one different but meaningful classification depending on the purpose of the investigation. To take a trivial card-games example, we naturally sort a hand of cards into suits when playing Bridge but into numbers when playing Rummy.

Another point to bear in mind is the possible existence of prior information. For example, in some situations the variables may be thought to have varying importance and this should clearly influence any analysis. However, most automatic clustering procedures find it difficult to take account of such information. As another example, we may have prior information regarding the likely number of groups and this will be helpful in finding a partition. However, Anderberg (1973, p. 24) points out that this prior information could be misleading if the data are a sample from some larger population and one (or more) of the groups happen to have been excluded from the sample. If the analyst still tries to find the full number of groups, then 'silly' clusters may be created.

In summary, we see that there are many practical problems involved in cluster analysis. The results of such an analysis will depend on a variety of considerations, such as which clustering method is chosen, which variables are measured, and which variables are thought to be important.

11.2 Visual approaches to finding a partition

If observations are taken on only two variables, the simplest, and probably the best, way of finding any natural groupings is to plot the data on a scatter diagram and examine the graph visually. Subjective judgement may then be used to identify any clusters which are present. Sometimes the clusters are obvious visually when they would be difficult to detect using an automatic algorithmic approach. This is particularly true when the clusters are curved or elongated as illustrated in Fig. 11.3(a). This is because most clustering

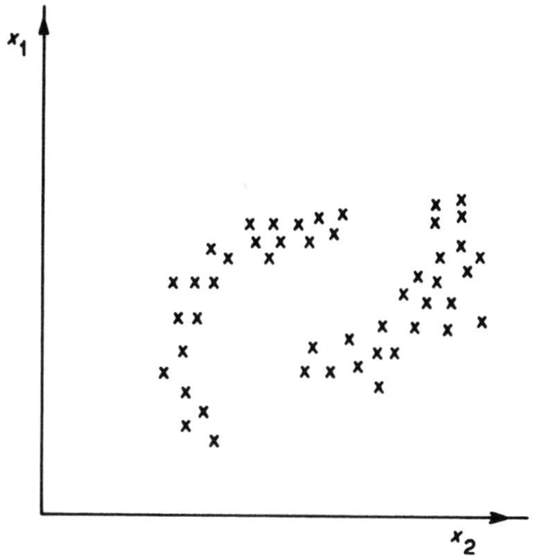

Figure 11.3(a). Two clusters which are not spherically shaped.

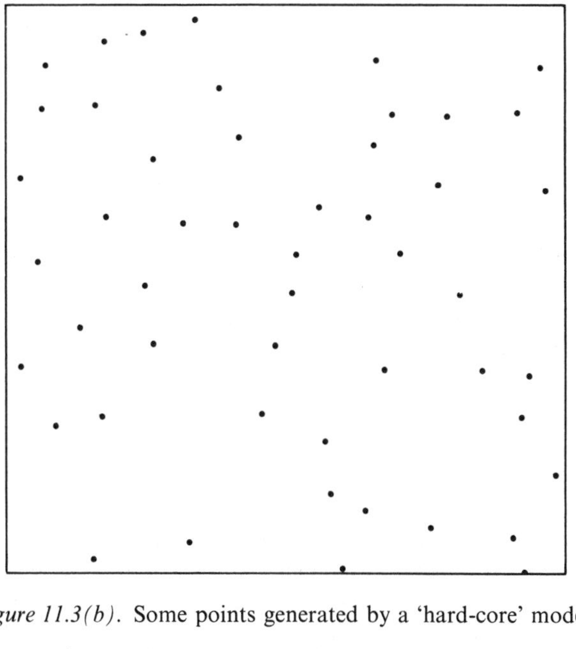

Figure 11.3(b). Some points generated by a 'hard-core' model.

procedures are quite good at detecting spherically shaped clusters but not so good at finding clusters of other shapes.

The human eye is not quite so good at detecting homogeneous scatter in a set of data, where there is no natural grouping and yet where an algorithmic approach might nevertheless detect spurious clusters. If the presence of clusters is doubtful, it may be advisable to carry out a formal test of the

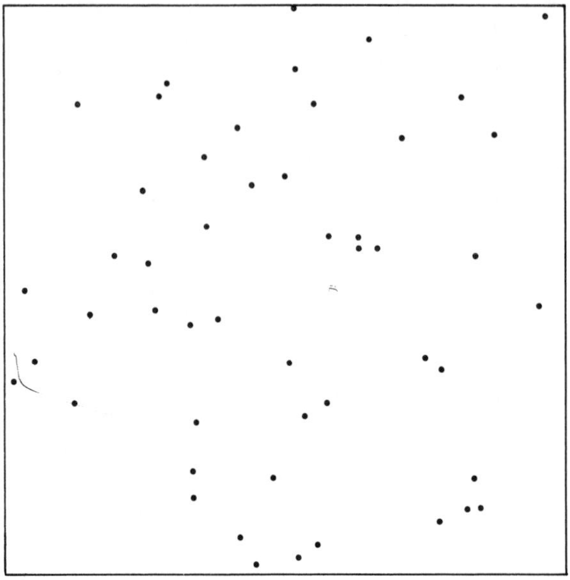

Figure 11.3(c). Some points generated by a planar Poisson process.

hypothesis that the points are randomly scattered over the plane (see Ripley, 1977). By 'random', we mean that the underlying model is a planar Poisson process. Many people might suspect that the points in Fig. 11.3(*b*) are 'random', but in fact they were generated by a model called a hard-core model, where points are not allowed to lie within a certain distance R from one another. So these points are *too* uniform. An example of 'random' data, generated by a planar Poisson process, is given in Fig. 11.3(*c*) and any perceived clustering is spurious. It is to be hoped that this example will help the reader in detecting homogeneous scatter.

With more than two variables, the graphical techniques described in Section 3.3 can be used to get a two-dimensional representation of the data and hence to find clusters visually. Of the direct methods, the use of Andrews curves seems most useful for identifying clusters. The indirect ordination methods can also be very useful. If 'distance' data have been recorded, we can try classical or ordinal scaling in two dimensions, plot the resulting map and look for clusters. If an $(n \times p)$ data matrix has been recorded, we can try principal component analysis. Then if it is found that the first two components 'explain' a 'large' proportion of the total variance, we can plot the scores on these two components for each individual so that we can again look for clusters visually. If *more* than two components are needed to give a satisfactory representation of the data, then principal component analysis will still be useful as it is probably safer and easier to try an algorithmic clustering approach on the scores of the first few important components rather than on the original data.

Visual approaches to finding a partition are perhaps not emphasized in the literature as much as they might be, possibly because of the subjective element involved. This, in our view, is a pity because visual approaches have an important role to play.

11.3 Hierarchical trees

Before we discuss hierarchical clustering procedures, it will be useful to consider more precisely what is meant by a hierarchical tree. The term was introduced in Section 11.1 and illustrated in Fig. 11.2.

A *tree* may be defined as a nested sequence of partitions of the individuals into g groups, where g varies from 1 to n, with the property that the partitions into g and into $(g + 1)$ groups are such that $(g - 1)$ of the groups are identical while the remaining individuals form one group in the first case and two groups in the second case. Thus, as defined concisely by Hartigan (1975, p. 11), a tree is a family of clusters for which any two clusters are either disjoint or one includes the other. The hierarchical structure is often represented by a two-dimensional diagram as in Figs. 11.2 and 11.4. This diagram is called a *tree diagram* or *dendrogram*. The tree is often presented 'upside down' as in Fig. 11.4, so that the 'branches' are at the bottom and the 'root' of the tree is at the top. However, when trees are produced by a computer it is often convenient to print them out so that the tree is on its side with the branches on the left. We should also point out that, although the individuals in Figs. 11.2 and 11.4 have been labelled consecutively to fit the tree arrangement, it will usually be necessary to re-order a pre-labelled set of individuals so that the branches of the tree do not cross over. It should also be stressed that the ordering of the branches is to some extent arbitrary. For example, in Fig. 11.4,

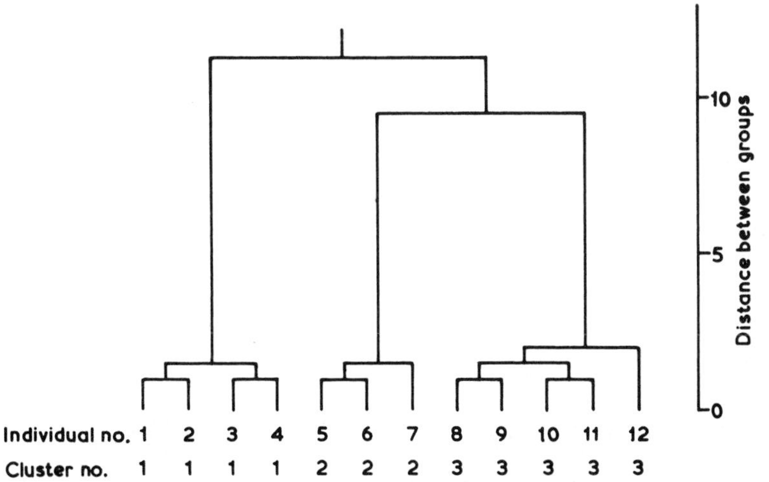

Figure 11.4. A tree diagram where three clusters are apparent.

individuals 1 and 2 could be interchanged with individuals 3 and 4 without making any difference to the tree structure. The ordering which appears depends on the particular algorithm used, and so no attempt should be made to interpret the order of the individuals which happens to occur in the displayed dendrogram.

A typical taxonomic tree in zoology would include super-families, families, species and subspecies in nested order.

Looking at Fig. 11.4, we see that there appears to be a natural partition into three clusters, as the individuals within the three clusters marked on the diagram are relatively close together, while the distances between the three clusters are relatively large. In contrast, there appears to be no natural partition in Fig. 11.2.

In Fig. 11.4, the reader will see a scale of distances labelled 'distances between groups'. Given a set of observed distances between all pairs of individuals, we shall see that there are many ways in which the distances between groups of individuals can be defined. Having chosen a suitable definition, the tree diagram is drawn so that two clusters unite at the appropriate derived distance. The tree diagram also implies a new set of distances between individuals which can be found from the distance level of the lowest link joining the two individuals in the tree diagram. These derived distances are sometimes called *threshold* distances. They satisfy the conditions of a metric dissimilarity coefficient, and in addition satisfy an inequality called the *ultrametric* inequality, which says that

$$d^*_{rs} \leq \max(d^*_{rt}, d^*_{ts}) \tag{11.1}$$

for all individuals r, s, t, where d^*_{rs} denotes the threshold distance between individuals r and s.

The above inequality implies that for any three individuals, either the three distances are the same, or more commonly that two are equal and the third is smaller. The reader should check that this holds for the threshold distances implied by Fig. 11.4 by taking three individuals at random.

A necessary and sufficient condition for an observed dissimilarity coefficient to be exactly represented by a dendrogram is that it should satisfy the ultrametric inequality. But most observed dissimilarities do not satisfy Equation (11.1), so that one could say that there is usually no genuine hierarchical structure. Nevertheless, a hierarchical clustering method tries to find a tree such that the derived ultrametric distances are in some sense as close as possible to the observed distances. This explains why a procedure for finding a tree from a given set of observed dissimilarities is sometimes called an *ultrametric transformation.*

11.4 Single-link clustering

Perhaps the most important method for finding a hierarchical tree is a method called the single-link method. This method may be defined as follows: For any threshold distance, d^*, the set of all individuals is divided into $g\,(\leq n)$ clusters such that individuals r and s are in the same cluster if there exists a chain of individuals $r, a, b, \ldots, q, s$, such that the observed dissimilarities in the chain, namely $d_{ra}, d_{ab}, \ldots, d_{qs}$, are all less than or equal to d^*. An equivalent definition involving graph theory is given by Jardine and Sibson (1971, p. 51).

There are many different numerical algorithms for actually finding the single-linkage tree. The easiest one to understand is the following:

(i) Start with n 'clusters', each containing just one individual.
(ii) Unite the two closest individuals, say r and s, into a single group, so that there are then $(n-1)$ groups.
(iii) The dissimilarity between this new group and any other individual t, is defined by $\min(d_{rt}, d_{st})$.
(iv) Unite the two closest 'groups', which will either be two individuals or one individual and the group formed in (ii).
(v) Construct new dissimilarities between the $(n-2)$ groups. Then continue to combine the groups so that at each stage the number of groups is reduced by one and the dissimilarity between any two groups is defined to be the dissimilarity between their closest members.

The property given in (v) above, that one compares groups by looking at their closest members, effectively means that only one link is required to join two groups. This explains the term 'single-link clustering'. It also explains an alternative name which is sometimes given to the method, namely the 'nearest-neighbour method'.

The treatment of ties is straightforward. If at any stage there are two dissimilarities which are equal but smaller than all the rest, then an arbitrary choice may be made as to which of the two closest pairs should be combined. The resulting dendrogram is not affected by which pair is initially selected.

The type of algorithm described above is called an *agglomerative* algorithm because it operates by a series of mergers, starting with n groups of just one individual and ending up with one group of n individuals.

Example 11.1. We will illustrate the above agglomerative algorithm by

Table 11.1 Dissimilarity matrix for five cars

Car	Car				
	1	2	3	4	5
1	—	.725	.925	.95	.935
2		—	.975	.94	.96
3			—	.955	.945
4				—	.69
5					—

Table 11.2 Revised dissimilarity matrix

	Car 1	Car 2	Car 3	Cars 4/5
Car 1	—	.725	.925	.935
Car 2		—	.975	.94
Car 3			—	.945
Cars 4/5				—

analysing the car data given in Table 10.1. On dividing by 200 and subtracting from 1, we get the dissimilarity matrix shown in Table 11.1.

First we look for the smallest dissimilarity. This is the value 0.69 between cars 4 and 5, so these 2 cars are joined to form a group at the threshold distance of 0.69. The dissimilarity between this group and the other cars can now be found. For example, the dissimilarity between car 1 and the group consisting of cars 4 and 5 is $\min(0.95, 0.935) = 0.935$. The revised dissimilarity matrix is as shown in Table 11.2.

The smallest dissimilarity is now that between cars 1 and 2, and so these two cars are merged at the second stage at a threshold distance of 0.725. The reader should complete this exercise and verify that at the third stage, cars 1/2 are merged with car 3 at the threshold distance of 0.925, and at the fourth stage all five cars join up at the threshold distance of 0.935. The resulting tree is shown in Fig. 11.5.

Of course, with only five 'individuals', the structure revealed by the tree can be seen fairly easily in the original similarity matrix, Table 10.1, where there are high frequencies joining cars 1 and 2 and cars 4 and 5. The rather artificial exercise given above is intended to demonstrate the agglomerative single-link algorithm for a simple case. □

Many alternative algorithms exist for finding a single-link tree. For example, we can find a *divisive* algorithm which operates by successive splitting of groups, starting with one group of n individuals and finishing with n groups

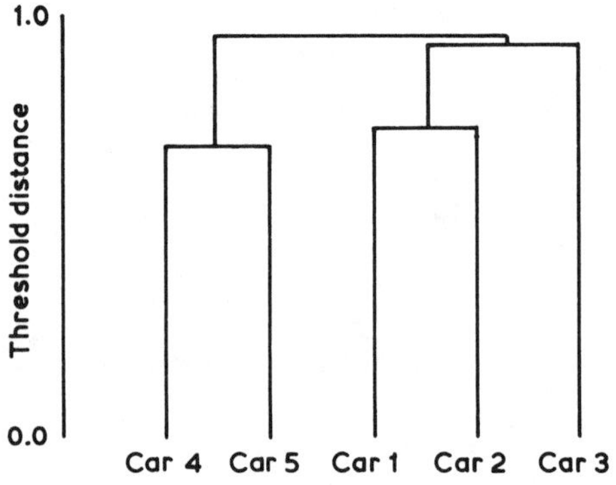

Figure 11.5. The hierarchical tree for the data in Table 11.1.

of one individual. This operates the opposite way round to an agglomerative algorithm where one has successive merging of groups. A single-link divisive algorithm is described by Hartigan (1975, p. 200) but it is computationally inefficient. If we start with one group of n individuals, it is easy to see that there are 2^{n-1} ways of making the first subdivision, and this large number makes divisive algorithms generally undesirable (except where several binary variables are measured and binary splitting is used, as will be described in the next section). In contrast, if we start with n groups of one individual there are only ${}^nC_2$ ways of combining two items at the first stage of an agglomerative algorithm.

An efficient single-link algorithm, which is neither agglomerative nor divisive, is described by Sibson (1973). An alternative algorithm is described by Gower and Ross (1969) which derives the single-link tree via a device called the *minimum spanning tree* (abbreviated MST), which is of interest in its own right. The MST is not a hierarchical tree, but rather is a network spanning all the points (or individuals) by a set of straight lines joining pairs of points whose lengths are equal to the appropriate interpoint dissimilarities. The MST is chosen so that:

(i) all pairs of points are connected;
(ii) the sum of the lengths of the straight lines is a minumum.

It is easy to see that the MST will not contain any closed loops and that each point is visited by at least one line. If there are ties in the dissimilarities, the MST may not be unique. Robin Sibson has suggested that it may be more incisive to define an invariant spanning graph, rather than an MST, as the set of links such that the observed dissimilarities are equal to the derived ultrametric distances. This graph is the same as the MST if there are no ties, but also gives a unique graph if there are ties, though it may then contain one or more closed loops and will not then be of minumum total length.

The minumum spanning tree for the data in Table 11.1 is shown in Fig. 11.6. Note that the angles between the straight lines are arbitrary and are chosen to spread out the branches of the tree, though the tree can also be superimposed

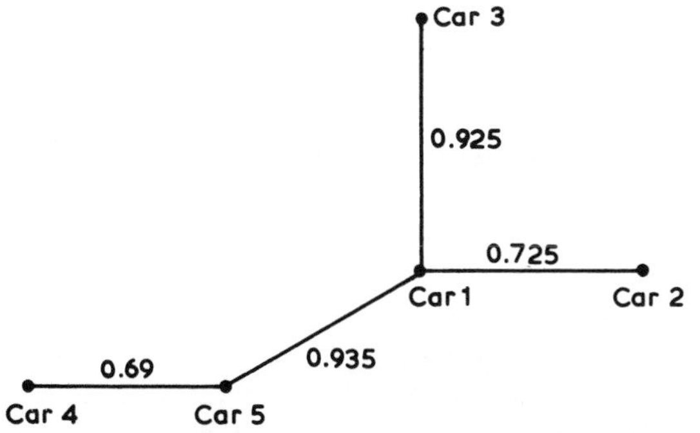

Figure 11.6. The minimum spanning tree for the data in Table 11.1.

onto some two-dimensional representation of the data. The minimum spanning tree should help in the identification of clusters, and also of outliers (see Barnett and Lewis, 1978, Section 6.2.8).

A little thought reveals that the minumum spanning tree of a set of data is closely related to the single-link tree. Thus the distances shown in Fig. 11.6 are the threshold distances which arise at successive stages in the agglomerative version of single-link clustering. Algorithms for finding the minimum spanning tree, and hence the single-link tree, are discussed by Gower and Ross (1969).

The advantages and disadvantages of single-link clustering will be discussed later in Section 11.6, after we have introduced some alternative clustering procedures.

11.5 Some other clustering procedures

Apart from single-link, many other clustering procedures are available for finding a hierarchical tree and/or a partition. In this section we will briefly introduce some of them, starting with hierarchical methods. Further details and references may be found in Anderberg (1973), Everitt (1974) and Hartigan (1975). In single-link clustering, the 'distance' between two groups of individuals is effectively defined to be the dissimilarity between their closest members. Several methods are apparently similar in spirit to single-link clustering in that they depend on alternative definitions of the 'distance' between two groups.

In the *complete-link* or *furthest-neighbour* method, the 'distance' between two groups is defined to be the dissimilarity between their most remote pair of individuals. In a sense this is the exact opposite of the single-link definition. In the *centroid* method, the 'distance' between two groups is defined to be the 'distance' between the group centroids (or group mean vectors). In the *group-average* method, the 'distance' between two groups is defined to be the average of the dissimilarities between all pairs of individuals such that there is one individual in each group. All the above methods can be implemented by means of an agglomerative algorithm.

Ward's hierarchical clustering method is based on within-group sums of squares rather than on ways of linking groups. An agglomerative algorithm is used. At each stage the number of groups is reduced by one, by combining the two groups which give the smallest possible increase in the total within-group sum of squares. Of course, when we start with n groups of just one individual, the total within-group sum of squares is zero.

Wishart's method, sometimes called *mode analysis*, searches for 'dense points' where k or more points (or individuals) are contained within a hypersphere of radius R. Starting with a 'small' value of R, the method looks at a hypersphere of radius R around each point and counts the number of other points within this hypersphere. If the number of points is at least k, then the centre point is called a *dense* point. The parameter R is gradually increased so that more and more points become dense, until all the points lie within a single hypersphere. If the parameter k is equal to 1, it is easy to see that the method is equivalent to single-link clustering.

One class of methods, called *binary splitting*, usually operates by means of a divisive algorithm, and is appropriate when most of the variables are binary attribute variables. Although divisive algorithms are generally less efficient than agglomerative algorithms, the reverse may hold with this sort of data. For example, in trying to classify a group of handicapped adults living at home, it is clearly important whether an individual can manage to climb the stairs or not, and this binary variable can easily be used to split the sample into two groups. A second binary variable, such as whether vision is impaired or not, can be used to further subdivide each group into two, giving four groups in all. A clear example is given by Jones (1979). *Association analysis* is a procedure of this type which is appropriate when one has several binary variables and wants to choose the one whose presence or absence will divide the sample into two subsets in the 'best' possible way. The *automatic interaction detector* (AID) method (e.g., Kass, 1975) is a related binary splitting method which is appropriate when one has a 'dependent' variable as well as several 'controlled' variables of a discrete type, possibly though not necessarily binary. At each stage the division of a group into two parts is defined by one of the controlled variables, with a subset of its possible categories defining one of the parts, and the remaining categories defining the other part.

The methods described so far are all essentially hierarchical clustering methods which produce a tree. It is of course possible to 'chop' a tree at an appropriate level in order to find a partition containing a given number of groups. But, as noted earlier, this may not necessarily be the best way of finding a partition, if a partition is required rather than a tree. Biologists tend to be mainly interested in finding a tree for different species, but in other applications a partition may be more useful.

A variety of partitioning techniques are available. These techniques usually involve a method for initiating a predetermined number of clusters. All the individuals are then allocated to one or other of the initiated clusters. Then some of the individuals are reallocated to other clusters. This iterative relocation of individuals continues until some criterion is optimized.

Several methods of initiating clusters have been proposed, including those given by E. M. L. Beale and J. MacQueen. For example, if we want a partition containing g clusters, we could choose g points to start the clusters which are the ones which are mutually furthest apart. The remaining points (or individuals) are then allocated to one of the initiated clusters, usually by simply finding the cluster to which they are nearest in some sense. The iterative relocation of individuals then continues until some criterion is optimized. One of the most popular criteria is to minimize Wilks's criterion, or equivalently the determinant of the within-group matrix of sums of squares and products.

We could go on to mention various other methods such as McQuitty's hierarchical clustering method and the 'flexible' scheme of Lance and

Williams. The latter approach embraces, as special cases, several methods which have already been introduced, including single-link and Ward's method. But we think we have already said enough to indicate the rich variety of methods available.

We have not attempted to describe the above methods in any detail as this would take too much space and would duplicate other work. The reader is referred to the books by Everitt (1974), Anderberg (1973) and Hartigan (1975), who not only describe the techniques in detail but also give the original references to the various methods.

11.5.1 *Method or algorithm?*

Having described a large number of clustering methods, it is now important to distinguish carefully between a clustering method and an algorithm for carrying it out, particularly for hierarchical clustering methods. Technically, a hierarchical clustering method maps a set of observed dissimilarity coefficients to a new set of dissimilarities which satisfy the ultrametric inequality and hence describe a hierarchical tree. It is important to realize that there may be several different algorithms for actually finding this mapping. For example, we have already seen that the single-link tree can be found by an agglomerative algorithm, by a divisive algorithm or via the minimum spanning tree. But some writers misleadingly describe the single-link method as an agglomerative method, not realizing that such terms as agglomerative and divisive apply, not to the cluster methods themselves, but to algorithms for implementing them. (See for example Jardine and Sibson, 1971, Section 6.2)

Of course, some methods are most conveniently defined in terms of a particular type of algorithm and then the distinction between 'method' and 'algorithm' is not so clear.

11.6 A comparison of procedures

With so many clustering methods available, it is difficult for the beginner to know how to choose the one which is most appropriate for a given set of conditions. Various factors need to be considered, including the theoretical questions discussed in the next subsection.

11.6.1 *Some desirable conditions for hierarchical clustering methods*

What mathematical conditions would we like a 'good' hierarchical clustering method to satisfy? This question is considered by Jardine and Sibson (1971, Section 9.3). A little thought will show that it is not easy to write down a set of strict mathematical conditions with which everyone will agree. Some of the conditions suggested by Jardine and Sibson, seem 'obvious', such as that the results produced by a method should not depend on the way the individuals are labelled. Others are rather more controversial than they might appear. For example, one might require a clustering procedure to be 'well defined', so that one always gets the same tree from the same set of observed dissimilarities. The difficulty with this condition arises when there are equal dissimilarities which are resolved in an arbitrary order during the sequential process of finding a tree. The single-link clustering method is 'well defined' but many other methods are not.

Perhaps the most crucial condition is the continuity condition suggested by Jardine and Sibson (1971, p. 86). Loosely speaking, this says that 'small' changes in the data should only produce 'small' changes in the resulting tree. Small changes in the dissimilarities can occur if the observations are subject to error, though one has to be careful to distinguish between changes in dissimilarities and changes in the original $(n \times p)$ data matrix if the dissimilarities are not observed directly. With some clustering methods, small changes in the data can produce large changes in the resulting tree, and this is clearly undesirable. An interesting example is given by Jardine and Sibson (1971, Section 7.4), where single-link clustering is shown to produce continuous changes in tree structure, whereas the complete-link and group-average methods give discontinuous changes.

Another important set of conditions are what Jardine and Sibson (1971, Section 9.4) call 'fitting-together' conditions. In particular, if one adds or subtracts just one individual from the original set one would like to see the tree structure change 'relatively little', though sometimes the classification will be changed in a non-trivial way. One possible requirement is that if one cuts off all the individuals in one branch of a tree, then the structure of the branch should be invariant when the clustering method is reapplied to the excised set of individuals.

The Jardine–Sibson conditions are discussed by Williams *et al.* (1971), who put an alternative point of view on some questions. In particular, they suggest that the continuity condition should be replaced by the alternative, vaguer condition that clusters should be relatively insensitive to outlying values or errors in the data.

11.6.2 *A comparison*

The choice of clustering method depends on a variety of factors including mathematical and computational considerations, and the type of data and variables which are involved. For example, binary splitting is appealing when several binary variables are observed, but not otherwise.

Single-link is the method of greatest mathematical appeal. It is the only hierarchical clustering method which satisfies all the conditions laid down by Jardine and Sibson (1971). In particular, methods like complete-link and the group-average method are unappealing because of their lack of continuity. Single-link also has computational advantages because the method is much quicker than many other methods and can even be implemented when there are several thousand individuals to compare. Single-link also gives solutions which are invariant in structure under a monotone transformation of the dissimilarity measure, so the method is good for data having ordinal significance only. (See Exercise 11.2)

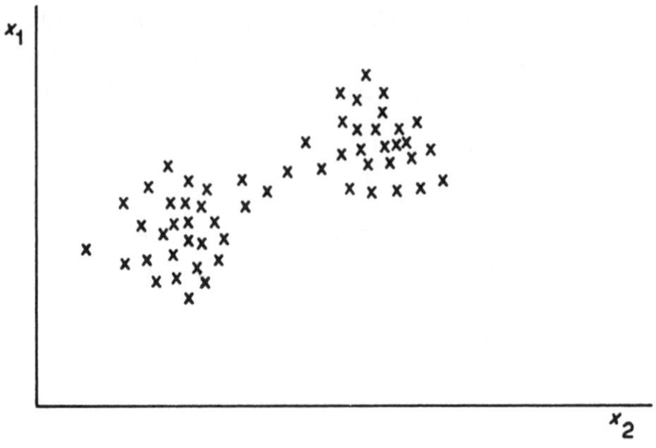

Figure 11.7. A set of two-dimensional data where 'chaining' will occur.

The main disadvantage of single-link is the so-called 'chaining' effect, which arises when apparently distinct clusters are united 'too soon' by a few intermediate points, as illustrated in Fig. 11.7 in two dimensions. Many of the alternative procedures were suggested in an effort to overcome this problem, but often merely succeed in introducing new problems such as a lack of continuity. The latter drawback is probably worse than chaining in many situations.

Although Jardine and Sibson (1971) recommend single-link as their preferred method, it is as well to remember that other writers have other preferences, fashioned no doubt by the particular problems which they have come across. For example, Marriott (1974) recommends Wishart's method as being 'effective and unlikely to give misleading results', provided that there is not a mixture of continuous and binary variates. The 'success' or 'failure' of different methods on particular sets of empirical data may have more influence than theoretical criteria in the selection of a method for a particular type of data.

Following up this last point, a number of empirical comparisons have been reported in the literature using artificially generated data, some of which are known to contain distinct groups. Various clustering methods are tried on each set of data but the results are inevitably inconsistent and no 'best buy' emerges. Everitt (1979) reviews several such empirical studies where single-link was on average rather less successful than some other methods, notably Ward's method. But single-link comes out rather better in the interesting study made by Everitt (1974, Chapter 4), who generated several different sets of bivariate normal data. The first set of data was from a single population so that there is just one group and no 'clusters' should be found. The remaining sets of data were constructed to contain two groups with different mean vectors. In some cases, equal sample sizes were taken and in other cases, unequal sample sizes. In some cases, the two variables had equal variances and were uncorrelated, giving spherically shaped clusters. In other cases, different variances, non-zero correlation and different covariance matrices for the two groups were tried. Everitt tried both partitioning and hierarchical techniques. What is remarkable about the results is that, even though distinct clusters have been built into the data, the results produced by the different methods are often so different from one another. The results as a whole show how much caution is necessary when trying to interpret the results of a cluster analysis.

Because some methods work well on certain types of data and not on others, the empirical studies emphasize the fact that most methods make implicit assumptions about the structure present in a sample. If these assumptions do not hold, a clustering method may *impose* a structure on the data rather than find what it is, so that spurious solutions may be found. For example, some methods tend to find spherically shaped 'clusters' whatever the shape of the true clusters, so that these methods are only 'good' when there happen to be clusters of this shape. As a second example, we note that if a partitioning technique is asked to find, say, five clusters, then it will do so even if the actual number of clusters is smaller or larger than five. As a third example, the partitioning technique, which minimizes the determinant of the matrix of within-group sums of squares and products, effectively assumes that the groups all have the same covariance matrix (i.e., have the same shape). When this is not the case, the method may produce 'wrong' answers. As a fourth example, the authors were recently asked to comment on the results of using Wishart's method on some data with the parameter $k = 4$. The analysts were surprised that a clear-cut cluster of two individuals, which was known beforehand to exist, was not found by the method. The reason is simple. By choosing the parameter k to be 4, the analyst is effectively saying that all groups must contain at least four individuals.

Bearing in mind the wide variety of practical situations, we find it rather hard to make general recommendations. Of the hierarchical clustering methods, single-link is probably to be preferred, although it may be worth trying more than one method. Indeed, it is sometimes suggested that the analyst should try several different cluster methods and, if they all reveal the same groupings, then one can be confident that natural groupings really do exist. But the empirical studies demonstrate that this is unlikely to happen unless the groups are spherically shaped and well separated. All in all, our preference is to avoid both partitioning and hierarchical techniques wherever possible and to use a visual approach as described in Section 11.2.

Exercises

11.1 Construct a tree (or dendrogram) using the single-link (or nearest-neighbour) method on the following data:

(*a*) The dissimilarity matrix calculated from the data in Exercise 10.1 using squared Euclidean distance.
(*b*) The distance matrix calculated from the data in Exercise 10.2.

(*c*) The dissimilarity matrix calculated from the data in Exercise 10.6 using the simple matching coefficient.

[Regarding part (*a*), note that with only two variables, it would be simpler in practice to plot the original data on a scatter diagram and look for clusters visually. The reader should do this in addition to the above, rather artificial exercise and see if the results agree. The reader could also construct the tree using the complete-link (or furthest-neighbour) method and see if the tree has the same structure as that arising from single-linkage.]

11.2 Suppose that d_1 and d_2 are two dissimilarity coefficients which are global-order equivalent. Suppose that two trees are constructed for a set of individuals using the single-link method on d_1 and then on d_2. Show that the trees have the same structure except that there may be a (possibly non-linear) transformation of the vertical scale measuring distances between groups.

References

Ahamad, B. (1967), 'An analysis of crimes by the method of principal components', *Appl. Statist.*, **16**, 17–35.

Anderberg, M. R. (1973), *Cluster Analysis for Applications*, New York: Academic Press.

Anderson, T. W. (1958), *An Introduction to Multivariate Analysis*, New York: Wiley, (2nd edn published 1984).

Anderson, T. W. (1963), 'Asymptotic theory for principal component analysis', *Ann. Math. Statist.*, **34**, 122–48.

Andrews, D. F. (1972), 'Plots of high dimensional data', *Biometrics*, **28**, 125–36.

Barnett, V. (1976), 'The ordering of multivariate data', *J. R. Statist. Soc.*, A, **139**, 318–54.

Barnett, V. and Lewis, T. (1978), *Outliers in Statistical Data*, Chichester: Wiley.

Bartlett, M. S. (1947), 'Multivariate analysis', *J. R. Statist. Soc.*, B, **9**, 176–97.

Beale, E. M. L., Kendall, M. G. and Mann, D. W. (1967), 'The discarding of variables in multivariate analysis', *Biometrika*, **54**, 357–66.

Bishop, Y. M. M., Fienberg, S. E., and Holland, P. W. (1975), *Discrete Multivariate Analysis*, Cambridge, Mass.: MIT Press.

Blackith, R. E. and Reyment, R. A. (1971), *Multivariate Morphometrics*, London: Academic Press (2nd edn published 1984 with Reyment, R. A., Blackith, R. E. and Campbell, N. A. as authors).

Box, G. E. P. (1949), 'A general distribution theory for a class of likelihood criteria', *Biometrika*, **36**, 317–46.

Carroll, J. D. and Chang, J. J. (1970), 'Analysis of individual differences in multidimensional scaling via an *N*-way generalisation of Eckert–Young decomposition', *Psychometrika*, **35**, 283–319.

Chakrapani, T. K. and Ehrenberg, A. S. C. (1979), 'Practical alternatives to factor analysis', Working paper, London Business School.

Chatfield, C. (1978), *Statistics for Technology* (2nd edn), London: Chapman and Hall, (3rd edn published 1983).

Chatfield, C. and Goodhardt, G. J. (1974), 'Results concerning brand choice', *J. Marketing Research*, **12**, 110–13.

Chernoff, H. (1973), 'Using faces to represent points in *k*-dimensional space graphically', *J. Amer. Statist. Ass.*, **68**, 361–8.

Cooley, W. W. and Lohnes, P. R. (1971), *Multivariate Data Analysis*. New York: Wiley.

Cormack, R. M. (1971), 'A review of classification', *J. R. Statist. Soc.*, A, **134**, 321–67.

Cox, D. R. (1970), *The Analysis of Binary Data*, London: Chapman and Hall.

Efron, B. and Morris, C. (1973*a*), 'Combining possibly related estimation problems' (with discussion), *J. R. Statist. Soc.*, B, **35**, 379—421.

Efron, B. and Morris, C. (1973*b*), 'Stein's estimation rule and its competitors – an empirical Bayes approach', *J. Amer. Statist. Ass.*, **68**, 117–30.

Efron, B. and Morris, C. (1975), 'Data analysis using Stein's estimator and its generalizations', *J. Amer. Statist. Ass.*, **70**, 311–319.

Ehrenberg, A. S. C. (1975), *Data Reduction*, Chichester: Wiley

Ehrenberg, A. S. C. (1977), 'Rudiments of numeracy' (with discussion), *J. R. Statist. Soc.*, A, **140**, 277–97.

Everitt, B. S. (1974), *Cluster Analysis*, London: Heinemann.

Everitt, B. S. (1977), *The Analysis of Contingency Tables*, London: Chapman and Hall.

Everitt, B. S. (1978), *Graphical Techniques in Multivariate Analysis*, London: Heinemann Educational Books.

Everitt, B. S. (1979), 'Unresolved problems in cluster analysis', *Biometrics*, **35**, 169–82.

Everitt, B. S. (1984), *An Introduction to Latent Variable Methods*, London: Chapman and Hall.

Gnanadesikan, R. (1977), *Methods for Statistical Data Analysis of Multivariate Observations*, New York: Wiley.

Gower, J. C. (1966), 'Some distance properties of latent root and vector methods used in multivariate analysis', *Biometrika*, **53**, 325–8.

Gower, J. C. and Ross, G. J. S. (1969), 'Minimum spanning trees and single linkage cluster analysis', *Appl. Statist.*, **18**, 54–64.

Granger, C. W. J. and Newbold, P. (1974), 'Spurious regressions in econometrics', *J. Econometrics*, **2**, 111–20.

Green, P. R. and Carmone, F. J. (1970), *Multidimensional Scaling and Related Techniques in Marketing Analysis*, Boston: Allyn and Bacon.

Greenacre, M. (1984), *Theory and Applications of Correspondence Analysis*, London: Academic Press.

Harman, H. H. (1976), *Modern Factor Analysis* (3rd edn), Chicago: University of Chicago Press.

Hartigan, J. A. (1975), *Clustering Algorithms*, New York: Wiley.

Hill, M. O. (1974), 'Correspondence analysis: A neglected multivariate method', *Appl. Statist.*, **23**, 340–54.

Hills, M. (1977), Book review, *Appl. Statist.*, **26**, 339–340.

Hodson, F. R., Kendall, D. G. and Tautu, P. (eds) (1971), *Mathematics in the Archaeological and Historical Sciences*, Edinburgh Univ. Press.

Ito, K. and Schull, W. J. (1964), 'On the robustness of the T_0^2 test in multivariate analysis of variance when variance-covariance matrices are not equal', *Biometrika*, **51**, 71–82.

James, W. and Stein, C. (1961), 'Estimation with quadratic loss', in *Proc. of 4th Berkeley Symposium*, Univ. of Calif. Press, **1**, 361–79.

Jardine, N. and Sibson, R. (1971), *Mathematical Taxonomy*, London: Wiley.

Jeffers, J. N. R. (1967), 'Two case studies in the application of principal component analysis', *Appl. Statist.*, **16**, 225–36.

Jeffers, J. N. R. (1977), 'Discriminant functions: A case study', *Bulletin in Applied Statistics* (BIAS), **4** (1), 24–38.

Jeffers, J. N. R. (1977), 'Lake district soils – a case study in cluster analysis', *Bulletin in Applied Statistics* (BIAS), **4** (2), 40–52.

Johnson, N. L. and Kotz, S. (1969), *Discrete Distributions*, Boston: Houghton Mifflin.

Johnson, N. L. and Kotz, S. (1972), *Distributions in Statistics: Continuous Multivariate Distributions*, New York: Wiley.

Joliffe, I. T. (1973), 'Discarding variables in a principal component analysis. II: Real data', *Appl. Statist.*, **22**, 21–31.

Jones, B. (1979), 'Cluster analysis of some social survey data', *Bulletin in Applied Statistics* (BIAS), **6**, 25–56.

Jöreskog, K. G., Klovan, J. E. and Reyment, R. A. (1976), *Methods in Geomathematics*, Vol. 1, *Geological Factor Analysis*, Amsterdam: Elsevier Scientific Publishing Co.

Kamen, J. M. (1970), 'Quick clustering', *J. Marketing Research*, **7**, 199–204.

Kass, G. V. (1975), 'Significance testing in automatic interaction detection', *Appl. Statist.*, **24**, 178–89.

Kendall, D. G. (1971), 'Construction of maps from "odd bits of information"', *Nature*, **231**, 158–9.

Kendall, D. G. (1975), 'The recovery of structure from fragmentary information', *Phil. Trans. of the Roy. Soc. of London*, **279**, 547–82.

Kendall, Sir Maurice (1957), *A Course in Multivariate Analysis*, London: Griffin.

Kendall, Sir Maurice (1975), *Multivariate Analysis*, London: Griffin, (2nd edn published 1980).

Khatri, C. G. (1966), 'A note on a MANOVA model applied to problems in growth curve', *Annals of the Inst. of Stat. Math.*, **18**, 75–86.

Kruskal, J. B. (1964*a*), 'Multidimensional scaling by optimizing goodness of fit to a nonmetric hypothesis', *Psychometrika*, **29**, 1–27.

Kruskal, J. B. (1964*b*), 'Nonmetric multidimensional scaling: A numerical method', *Psychometrika*, **29**, 115–29.

Kruskal, J. B. and Wish, M. (1978), *Multidimensional Scaling*, London: Sage University Paper Series on Quantitative Applications in the Social Sciences.

Lachenbruch, P. A. and Goldstein, M. (1979), 'Discriminant analysis', *Biometrics*, **35**, 69–86.

Lawley, D. N. and Maxwell, A. E. (1971), *Factor Analysis as a Statistical Method* (2nd edn), London: Butterworth.

Lee, Y. S. (1971), 'Asymptotic formulae for the distribution of a multivariate test statistic: Power comparisons of certain multivariate tests', *Biometrika*, **58**, 647–51.

Lindley, D. V. and Smith, A. F. M. (1972), 'Bayes estimates for the linear model' (with discussion), *J. R. Statist. Soc.*, B, **34**, 1–41.

Lingoes, J. C. and Roskam, E. E. (1973), 'A mathematical and empirical analysis of two multidimensional scaling algorithms', *Psychometrika*, **38**, Monograph supplement, 1–93.

Mardia, K. V., Kent, J. T. and Bibby, J. M. (1979), *Multivariate Analysis*, London: Academic Press.

Marriott, F. H. C. (1974), *The Interpretation of Multiple Observations*, London: Academic Press.

Maxwell, A. E. (1977), *Multivariate Analysis in Behavioural Research*, London: Chapman and Hall.

Morrison, D. F. (1976), *Multivariate Statistical Methods* (2nd edn), New York: McGraw-Hill.

Naus, J. I. (1975) *Data Quality Control and Editing*, New York: Marcel Dekker.

Nelder, J. A. and Wedderburn, R. W. M. (1972), 'Generalized linear models', *J. R. Statist. Soc.*, A, **135**, 370–84.

Olsen, C. L. (1974), 'Comparative robustness of six tests in multivariate analysis of variance', *J. Amer. Statist. Ass.*, **69**, 894–908.

O'Muircheartaigh, C. A. and Payne, C. (eds) (1977), *The Analysis of Survey Data*, Vol. 1, *Exploring Data Structures*, Chichester: Wiley.

Pillai, K. C. S. and Jayachandran, K. (1967), 'Power comparisons of tests of two multivariate hypotheses based on four criteria', *Biometrika*, **54**, 195–210.

Plackett, R. L. (1974), *The Analysis of Categorical Data*, London: Griffin.

Rao, C. R. (1951), 'An asymptotic expansion of the distribution of Wilks' Criterion', *Bull. Int. Stat. Inst.*, **33(2)**, 177–180.

Rao, C. R. (1973), *Linear Statistical Inference and its Applications* (2nd edn), New York: Wiley.

Ripley, B. D. (1977), 'Modelling spatial patterns' (with discussion), *J. R. Statist. Soc.*, B, **39**, 172–212.

Romney, A. K., Shepard, R. N. and Nerlove, S. B. (eds) (1972), *Multidimensional Scaling*, Vol. II, *Applications*, New York: Seminar Press.

Roy, S. N. and Bose, R. C. (1953), 'Simultaneous confidence interval estimation', *Ann. Math. Statist.*, **24**, 513–36.

Roy, S. N., Gnanadesikan, R. and Srivastava, J. N. (1971), *Analysis and Design of Certain Quantitative Multiresponse Experiments*, Oxford: Pergamon Press.

Schatzoff, M. (1966), 'Sensitivity comparisons among tests of the general linear hypothesis', *J. Amer. Statist. Ass.*, **61**, 415–35.

Scheffé, H. (1959), *The Analysis of Variance*, New York: Wiley.

Seal, H. L. (1964), *Multivariate Statistical Analysis for Biologists*, London: Methuen.

Sibson, R. (1972), 'Order invariant methods for data analysis', *J. R. Statist. Soc.*, B, **34**, 311–49.

Sibson, R. (1973), 'SLINK: An optimally efficient algorithm for the single-link cluster method', *Computer J.*, **16**, 30–4.

Sibson, R. (1979), 'Studies in the robustness of multidimensional scaling: Perturbational analysis of classical scaling', *J. R. Statist. Soc.*, B., **41**, 217–29.

Sokal, R. R. and Sneath, P. H. A. (1963), *Principles of Numerical Taxonomy*, London: Freeman.

Sprent, P. (1969), *Models in Regression*, London: Methuen.

Springall, A. (1978), 'A review of multidimensional scaling', *Bulletin in Applied Statistics* (BIAS), **5**, 146–92.

Stein, C. (1962), 'Confidence sets for the mean of a multivariate normal distribution' (with discussion), *J. R. Statist. Soc.*, B, **24**, 265–96.

Torgerson, W. S. (1952), 'Multidimensional scaling. I. Theory and method', *Psychometrika*, **17**, 401–19.

Torgerson, W. S. (1958), *Theory and Methods of Scaling*, New York: Wiley.

Walker, M. A. (1967), 'Some critical comments on "An analysis of crimes by the method of principal components" by B. Ahamad', *Appl. Statist.*, **16**, 36–9.

Welford, B. P. (1962), 'Note on a method for calculating corrected sums of squares and products', *Technometrics*, **4**, 419–20.

Williams, W. T., Lance, G. N., Dale, M. B. and Clifford, H. T. (1971), 'Controversy concerning the criteria for taxonometric strategies', *Comp. J.*, **14**, 162–5.

Answers to exercises

Chapter 1

1.2

$$\begin{bmatrix} \frac{1}{2} & 0 & 0 & 0 \\ 0 & \frac{1}{4} & 0 & 0 \\ 0 & 0 & \frac{5}{14} & -\frac{1}{14} \\ 0 & 0 & -\frac{1}{14} & \frac{3}{14} \end{bmatrix}$$

1.3

$$\begin{bmatrix} \cos\theta & -\sin\theta \\ \sin\theta & \cos\theta \end{bmatrix}$$

1.4

$$\begin{bmatrix} 1/\sqrt{3} & 1/\sqrt{3} & 1/\sqrt{3} \\ 1/\sqrt{2} & -1/\sqrt{2} & 0 \\ 1/\sqrt{6} & 1/\sqrt{6} & -2/\sqrt{6} \end{bmatrix}$$ is one possibility.

1.5 Substract the first row from each other row. Then subtract each column from the first column to get $|A|.|A-\lambda I|$ is of the same form when the constant $(1-\lambda)^p$ is taken outside and every term in the matrix is divided by $(1-\lambda)$.

1.6 Roots are $\lambda=4$, once, and $\lambda=1$, twice. First eigenvector is $(1/\sqrt{3})[1, 1, 1]$. Second and third eigenvectors (which are not unique) could be $(1/\sqrt{2})[1, -1, 0]$ and $(1/\sqrt{6})[1, 1, -2]$

1.7 A is of order $(n \times n)$ with diagonal terms $(n-1)/n$ and off-diagonal terms $-1/n$. It is positive semidefinite with rank $(n-1)$.

1.8 rank $= 2$.

1.9 $A=(1+r)\begin{bmatrix} \frac{1}{2} & \frac{1}{2} \\ \frac{1}{2} & \frac{1}{2} \end{bmatrix}+(1-r)\begin{bmatrix} \frac{1}{2} & -\frac{1}{2} \\ -\frac{1}{2} & \frac{1}{2} \end{bmatrix}$

Chapter 2

2.1 The conditional distribution of X given $Y=\frac{1}{2}$, $Z=1$, is the same as the marginal distribution of X.

2.2 $k=3$. Marginal p.d.f. of X is $3x^2$ for $0<x<1$, and that of Y is $3(1-y^2)/2$ for $0<y<1$. Their product is *not* equal to the joint p.d.f. so X and Y are *not* independent.

2.3 $$\begin{aligned}\text{Var}(aX+bY) &= a^2\text{Var}(X)+2ab\text{Cov}(X, Y)+b^2\text{Var}(Y) \\ &\geq 0 \quad \text{for every } a, b.\end{aligned}$$

Put $a=\sqrt{[\mathrm{Var}(Y)]}$ and $b=\sqrt{[\mathrm{Var}(X)]}$, which gives $\mathrm{Cov}(X, Y) \geq -\sqrt{[\mathrm{Var}(X)\mathrm{Var}(Y)]}$. Hence $\rho \geq -1$. Put $b=-\sqrt{[\mathrm{Var}(X)]}$ to obtain $\rho \leq +1$. In the example $f(x, y)$ is symmetric about both the x- and y-axes, and so we must have $\rho = 0$. But $f(x, y)$ does not factorize into a function of x multiplied by a function of y, so X and Y are not independent.

2.4 $\Sigma_Y = \begin{bmatrix} 3 & 0 & 0 \\ 0 & 2 & 1 \\ 0 & 1 & 2 \end{bmatrix}$, $\mathrm{P}_Y = \begin{bmatrix} 1 & 0 & 0 \\ 0 & 1 & \frac{1}{2} \\ 0 & \frac{1}{2} & 1 \end{bmatrix}$.

2.5
$$E[A^T\mathbf{X}] = A^T\boldsymbol{\mu}$$
$$\mathrm{Var}[A^T\mathbf{X}] = E[(A^T\mathbf{X} - A^T\boldsymbol{\mu})(A^T\mathbf{X} - A^T\boldsymbol{\mu})^T] = A^T\Sigma A$$

2.6 Σ is positive semidefinite and $\mathrm{P} = D^{-1}\Sigma D^{-1}$ where D is non-singular. $\mathbf{a}^T P\mathbf{a} = \mathbf{a}^T D^{-1}\Sigma D^{-1}\mathbf{a} = \mathbf{b}^T\Sigma\mathbf{b} \geq 0$ for every $\mathbf{a}$, where $\mathbf{b} = D^{-1}\mathbf{a}$. Σ and P are positive definite if the variables are linearly independent in the algebraic sense.

2.8 $\Sigma = \begin{bmatrix} \sigma_1^2 & \rho\sigma_1\sigma_2 \\ \rho\sigma_1\sigma_2 & \sigma_2^2 \end{bmatrix}$ $\mathrm{P} = \begin{bmatrix} 1 & \rho \\ \rho & 1 \end{bmatrix}$

$$\Sigma^{-1} = \frac{1}{(1-\rho^2)} \begin{bmatrix} 1/\sigma_1^2 & -\rho/\sigma_1\sigma_2 \\ -\rho/\sigma_1\sigma_2 & 1/\sigma_2^2 \end{bmatrix}$$

(*a*) X_1, X_2 are independent. (*b*), (*c*) The distribution is degenerate. There is an exact linear relationship between X_1 and X_2 with positive and negative coefficients respectively.

Chapter 3

3.2
$$\sum(y_r - \bar{y})^2 = \mathbf{a}^T\{\sum(\mathbf{x}_r - \bar{\mathbf{x}})(\mathbf{x}_r - \bar{\mathbf{x}})^T\}\mathbf{a} = \mathbf{a}^T X^T X\mathbf{a} = (n-1)\mathbf{a}^T S\mathbf{a} \geq 0 \quad \text{for every } \mathbf{a}.$$

Chapter 4

4.2 $\lambda_1 = 10, \lambda_2 = 3$. $\mathbf{a}_1^T = (1/\sqrt{7})[\sqrt{6}, 1]$. $\mathbf{a}_2^T = (1/\sqrt{7})[1, -\sqrt{6}]$.
$\mathrm{P} = \begin{bmatrix} 1 & 1/\sqrt{6} \\ 1/\sqrt{6} & 1 \end{bmatrix}$ has eigenvalues $1 \pm 1/\sqrt{6}$, with
$\mathbf{a}_1^T = (1/\sqrt{2})[1, 1]$, $\mathbf{a}_2^T = (1/\sqrt{2})[1, -1]$.
For Σ, $Y_1 = (\sqrt{6}X_1 + X_2)/\sqrt{7}$. For P, $Y_1^* = (X_1^* + X_2^*)/\sqrt{2} = (X_1/3 + X_2/2)/\sqrt{2}$, and the coefficients of X_1 and X_2 are not in the same ratio as in Y_1.

4.3 Eigenvalues of $7S$ are 400 and 100. $\mathrm{r} = 0.58$.

4.4 $\lambda_1 = 5/2, \quad \lambda_2 = \lambda_3 = \lambda_4 = 1/2. \quad \mathbf{a}_1^T = \frac{1}{2}[1, 1, 1, 1]$.

$\Sigma^* = \begin{bmatrix} 4 & 1 & 1 & 1 \\ 1 & 1 & \frac{1}{2} & \frac{1}{2} \\ 1 & \frac{1}{2} & 1 & \frac{1}{2} \\ 1 & \frac{1}{2} & \frac{1}{2} & 1 \end{bmatrix}$ $\quad \lambda_1 = 5, \lambda_2 = 1, \lambda_3 = \lambda_4 = \frac{1}{2}$, $\quad \mathbf{a}_1^{*T} = [3,1,1,1]/\sqrt{12}$.

Scaled back, $\mathbf{a}_1^{*T}$ becomes proportional to $[3/2,1,1,1]$, not $[1,1,1,1]$.

4.5 $\lambda_1 = 6, \lambda_2 = 3, \lambda_3 = 2$. $\mathbf{a}_1^T = [1,1,2]/\sqrt{6}$.

$\mathbf{a}_2^T = [1,1,-1]/\sqrt{3}$. $\mathbf{a}_3^T = [1,-1,0]/\sqrt{2}$. Proportion is 9/11.
$\mathbf{Y} \sim N_3(\mathbf{0}, \Lambda)$ where $\Lambda = \begin{bmatrix} 6 & 0 & 0 \\ 0 & 3 & 0 \\ 0 & 0 & 2 \end{bmatrix}$

4.6 $Y_1 = (X_1 + X_2)/\sqrt{2}$. $Y_2 = (X_1 - X_2)/\sqrt{2}$ if $\rho > 0$. Y_1, Y_2 are reversed if $\rho < 0$.

4.7 $S = \sum_{i=1}^{4} \lambda_i \mathbf{a}_i \mathbf{a}_i^T$ as in Exercise 4.1. See Section 4.3.4 for last part.

Chapter 6.

6.1 (i) $\Sigma = \begin{bmatrix} 5 & 3 & 3 \\ 3 & 3 & -2 \\ 3 & -2 & 14 \end{bmatrix}$ of rank 3. (ii) $\Sigma = \begin{bmatrix} 5 & 3 & 3 \\ 3 & 2 & 1 \\ 3 & 1 & 5 \end{bmatrix}$ of rank 2.

6.2 (i) $\lambda_1 = 12, \lambda_2 = 4, \lambda_3 = 3$; $B = C\Lambda^{1/2} = \begin{bmatrix} 2/\sqrt{6} & 0 & 1/\sqrt{3} \\ -1/\sqrt{6} & 1/\sqrt{2} & 1/\sqrt{3} \\ -1/\sqrt{6} & -1/\sqrt{2} & 1/\sqrt{3} \end{bmatrix} \Lambda^{1/2}$.

Check that $\Sigma = \sum_{i=1}^{3} \lambda_i \mathbf{c}_i \mathbf{c}_i^T$ [The spectral decomposition of Σ]

(ii) $\lambda_1 = 12, \lambda_2 = 4, \lambda_3 = 0$; $B = C\Lambda^{1/2} = \begin{bmatrix} 2\sqrt{2} & 0 \\ -\sqrt{2} & \sqrt{2} \\ -\sqrt{2} & -\sqrt{2} \end{bmatrix}$

Structural relation from $\mathbf{c}_3 \Rightarrow X_1 + X_2 + X_3 =$ constant.

6.3 For Σ in 6.2(i), $k_{11} = 3$, $k_{21} = k_{31} = -1$, $k_{22} = k_{33} = 2$, $k_{32} = 0$.
For Σ in 6.2(ii), $k_{11} = 2\sqrt{2}$, $k_{21} = k_{31} = k_{32} = -\sqrt{2}$, $k_{22} = \sqrt{2}$, $k_{33} = 0$. X_1, X_2 and X_3 are linear compounds of Z_1 and Z_2 only.

6.4 $\mathrm{Var}\left(X_1 \middle| \begin{bmatrix} X_2 \\ X_3 \end{bmatrix}\right) = 1/38, 0, 6, 0$ in the four examples (degenerate when zero)

$\mathrm{Var}\left(\begin{bmatrix} X_1 \\ X_2 \end{bmatrix} \middle| X_3\right) = (1/14)\begin{bmatrix} 61 & 48 \\ 48 & 38 \end{bmatrix}, (1/5)\begin{bmatrix} 16 & 12 \\ 12 & 9 \end{bmatrix},$

$(12/5)\begin{bmatrix} 3 & -1 \\ -1 & 2 \end{bmatrix}, \begin{bmatrix} 4 & -4 \\ -4 & 4 \end{bmatrix}$ (degenerate).

6.6 If $\mu_r = \begin{cases} \mu + \tau_1 & r = 1,2,3 \\ \mu + \tau_2 & r = 4,5,6 \end{cases}$, $\mathbf{V}_1 \sim N_3(\sqrt{6}\boldsymbol{\mu}, \Sigma)$, $\mathbf{V}_2 \sim N_3\left(\frac{\sqrt{6}}{2}(\tau_1 - \tau_2), \Sigma\right)$,

$\mathbf{V}_r \sim N_3(\mathbf{0}, \Sigma)$, $r = 3,4,5,6$.

6.7 *Hint*: Compare the univariate analyses of variance for the linear compounds $\mathbf{c}^T\mathbf{X}$ for $\mathbf{c}^T = [1,0,0]$, $[0,1,0]$ and $[1,1,0]$.

6.9 The $\mathbf{d}_r$ are orthonormal so that the $\mathbf{V}_r$ are independently distributed $N_3(\boldsymbol{\nu}_r, \Sigma)$ where the $\boldsymbol{\nu}_r$ are given in the answers to Exercise 6.6. Hence $4S = \sum_{r=3}^{6} \mathbf{V}_r \mathbf{V}_r^T \sim W_3(4, \Sigma)$ independently of $\mathbf{V}_2$. The result follows using Equation (6.25).

Chapter 7

7.2 (i) $\mathcal{T}^2 = 13.37$; $\mathcal{F}(3, 6) = 3.34$.
(ii) $\mathcal{T}^2 = 25.79$; $\mathcal{F}(2, 7) = 11.28$.

7.3 (*a*) Use test in Equation (7.22).

(*b*) Under $H_0: \dfrac{\mu_{i+2}-\mu_{i+1}}{t_{i+2}-t_{i+1}} = \dfrac{\mu_{i+1}-\mu_i}{t_{i+1}-t_i}$, $i = 1, 2, \ldots, (p-2)$.

Use $\bar{\mathbf{x}}$ and S calculated from male patients only and take

$$C^{\mathrm{T}} = \begin{bmatrix} 12 & -15 & 3 & 0 & 0 \\ 0 & 3 & -5 & 2 & 0 \\ 0 & 0 & 2 & -3 & 1 \end{bmatrix}$$

$H_0: C^{\mathrm{T}}\boldsymbol{\mu} = \mathbf{0}$ is tested using Equation (7.22). $\mathscr{T}^2 = 11.42$; $\mathscr{F}(3, 2) = 1.90$.

7.4 $\mathscr{T}^2 = 3.89$; $\mathscr{F}(2, 12) = 1.79$. A test of flatness is not relevant.

7.5 $\mathscr{T}^2 = 37.96$; $\mathscr{F}(5, 4) = 3.80$.

Linear discriminant ($\mathbf{L} = \mathbf{a}^* = S^{-1}\mathbf{d}$) is

$$U = 0.7491X_1 - 2.0308X_2 - 0.5351X_3 + 2.3423X_4 - 0.2175X_5$$

All patients in the samples would have been correctly diagnosed with the rule: If $U < 3.456$, allocate to group 1; otherwise to group 2. Information about prior probabilities would reduce the critical value by ln 9.

Chapter 8

8.1 (i) $\mathrm{SSPM(Nitrogen)} = N = \begin{bmatrix} 386.75 & 73.75 \\ 73.75 & 14.75 \end{bmatrix}$

$$\mathrm{SSPM(Residual)} = R = \begin{bmatrix} 93.00 & -50.00 \\ -50.00 & 33.00 \end{bmatrix}$$

(ii) Using tables of orthogonal polynomials, linear nitrogen SS for Y_1 is $174^2/80$ and for Y_2 is $34^2/80$. The product term is calculated as $(174 \times 34)/80$ giving

$\mathrm{SSPM(Lin.}\ N) = \mathrm{lin.}\ N = \begin{bmatrix} 378.45 & 73.95 \\ 73.95 & 14.45 \end{bmatrix}$; non-lin. N is obtained by subtraction

(iii) (*a*) 0.0255 (*b*) 0.6669 (*c*) 0.0261.

Non-linear effect is not significant.

8.5 (i) $\theta_1 = 37.7899$, $\theta_2 = 0.01235$; dimensionality is not significantly different from 1. The first canonical variate is $Z_1 = 0.83Y_1 + 1.32Y_2$

(ii) $Z_2 = -0.09Y_1 + 0.46Y_2 = 2.79$.

(iii) $\theta_1 = 37.3070$, $\theta_2 = 0$; $Z_1 = 0.8293Y_1 + 1.3219Y_2$

(iv) $Z_2 = -0.0904Y_1 + 0.4625Y_2 = 10.9785$

8.6 $\Lambda_{T\times S} = 0.8228$ (not significant); $\Lambda_T = 0.4050$, $\mathscr{F}(4, 22) = 3.14$. $\Lambda_S = 0.3427$, $\mathscr{F}(2, 11) = 10.5$

Chapter 9

9.1 $U_1 = 0.4794$ (chest circumference) $+ 0.1714$(MUAC)

$V_1 = 0.1124$(height) $+ 0.3552$(age)

$\hat{u}_r - 31.34 = 0.99(v_r - 17.39)$, $\hat{v}_r - 17.39 = 0.99(u_r - 31.34)$.

$\nu_1 = \rho_1^2 = 0.9805$, $\nu_2 = \rho_2^2 = 0.6792$

Test (9.35) for dimensionality 1: $-2.5\ln(1-\nu_2) = 2.84 < \chi^2_{.05}(1)$

9.2 $\hat{\mathrm{B}} = \begin{bmatrix} 0.2169 & 0.0720 \\ 0.3274 & 0.4431 \end{bmatrix}$, $\mathbf{a}_1 = \begin{bmatrix} 0.5010 \\ 0.2162 \end{bmatrix}$, $\mathbf{b}_1 = \begin{bmatrix} 0.1308 \\ 0.2735 \end{bmatrix}$

9.3 (*a*) $\Lambda_{2.1,2,(4-1)} = \Lambda_3/\Lambda_1 = 0.004313/0.4569 = 0.009440 (\Lambda_1 = 1.06/(1.26 + 1.06))$

$$\frac{1-\Lambda^{1/2}}{\Lambda^{1/2}}\frac{3-2+1}{2} = 9.29 > F_{.05}(4, 4)$$

(*b*) $\Lambda_{1.2,2,2} = \Lambda_3/\Lambda_2 = 0.004313/0.09656 = 0.04467$ (Λ_2 using X_2 and X_3)

$$\frac{1-\Lambda}{\Lambda}\frac{2-1+1}{1} = 42.78 < F_{.05}(1, 2).$$

9.4 $\Lambda_{1.2,2,2} = \Lambda_3/\Lambda_2 = 0.004313/0.1483 = 0.02909$ (Λ_2 using X_1 and X_3).

9.6 *Hint:* Use Equation (1.4) to rewrite $|R|/|R_{33}|$.

9.7 Obtain $13S^* = 13S_{3.1} = \begin{bmatrix} 20.1728 & 5.0706 & 74.6324 \\ 5.0706 & 8.9412 & 24.7647 \\ 74.6324 & 24.7647 & 344.0588 \end{bmatrix}$

$\bar{\mathbf{d}}^* = \begin{bmatrix} 2.2103 \\ 1.4706 \\ 6.3824 \end{bmatrix}$ where, for example, $20.1728 = 31.62 - (27.90)^2/68.00$

$5.0706 = 18.20 - (27.90)(32.00)/68.00$

$2.2103 = 1.8 - (27.90/68.00)(-1.0)$.

The resulting discriminant function is

$$0.2187(\text{chest circumference})^* + 0.1505(\text{MUAC})^* - 0.0397(\text{height})^*$$

where the asterisk denotes the variable adjusted for age. Records for an individual whose sex was not recorded are adjusted for age to the grand mean age and the discriminant function is used to allocate the individual to one sex group or the other. This is not the same as the discriminant in Example 9.6.

Chapter 10

10.1 Both 5 × 5 matrices have zeroes on the diagonals. As an example, d_{12} is 1 for (*a*) and 1 for (*b*), while d_{13} is 29 for (*a*) and 7 for (*b*).

10.2 As an example, $d_{34} = 6$.

10.3 Both d_1 and d_2 are obviously DC's. But the proof of the metric inequality is tricky and will not be given here. d_3 is a DC, but find a counter example to show that it does no satisfy the metric inequality. d_4 is not a DC as $d_{rr} \neq 0$.

10.6 As an example, for (*a*), $d_{45} = 3/5$; while for (*b*), $d_{45} = 3/4$.

Chapter 11

11.1 As an example, in part (*c*) using an agglomerative algorithm, we find individuals 2 and 5 join up first at threshold distance zero; then 1 and 3 join up at threshold 1/5; these two pairs join up at threshold 2/5; finally individual 4 joins up at threshold 3/5.

Name index

Subject index